# Mechanical Properties of MAX Phases

David J. Fisher

Published by **Materials Research Forum LLC**
Millersville, PA 17551, USA

Published as part of the book series
**Materials Research Foundations**
Volume 97 (2021)
ISSN 2471-8890 (Print)
ISSN 2471-8904 (Online)

Print ISBN 978-1-64490-126-7
ePDF ISBN 978-1-64490-127-4

Distributed worldwide by

**Materials Research Forum LLC**
105 Springdale Lane
Millersville, PA 17551
USA
http://www.mrforum.com

Printed in the United States of America
10 9 8 7 6 5 4 3 2 1

# Table of Contents

## Introduction

These are materials which generally comprise three principal elements, perhaps with others in solid solution, and have great potential importance in fields ranging from the electronic to the mechanical. The present work is concerned only with the mechanical properties, and the primary attraction here is that these materials exhibit properties which can simultaneously be both metallic and ceramic in nature.

Their name arises from the designations of the three elements, with M being a transition metal, A usually being another metal or metalloid (cadmium, aluminium, tin, lead, gallium, indium, thallium, silicon, germanium, phosphorus, arsenic, sulfur) and X being carbon or nitrogen. In recent years, a similar class of materials has attracted interest. In these, the third element is boron. These are known as MAB phases, and a number of them are included in the present work, for comparison purposes. The M, A and X elements are present in simple numerical ratios, and the various types of MAX phases are conveniently designated by that ratio. The 211-type phases are of the form, $M_2AX$. This is one of the largest groups and these phases possess a nanolaminate structure within which blocs of a $M_2X$, carbide or nitride, are separated by monatomic layers of A. Due to a high crystal symmetry, the structure is conveniently defined merely by the *a* and *c* lattice constants and the interplanar separation of the M and X layers. As well as the layering of the structure, the nature of the bonding, covalent or ionic or metallic, is important. Another useful feature is that the chemistry of a MAX phase can be easily changed without affecting the structure.

These nanolaminates generally consist of hexagonal carbide or nitride blocs and planar A atomic sheets with a zig-zag stacking along the z-axis. It is this layered structure which imparts to the phases an unusual combination of ceramic and metallic properties. Among the ceramic properties are rigidity, oxidation-resistance and high-temperature strength. Among the metallic properties are machinability, thermal-shock resistance, damage-tolerance and good transport properties. The MAX phases are relatively soft and the Vickers hardness values of polycrystalline samples tend to range from 2 to 8GPa, making them generally softer than structural ceramics but harder than metals.

The essentially two-dimensional nature of these phases means that the dislocations also tend to be confined to the two dimensions of the basal planes upon which the dislocations glide. This facilitates the general understanding of their deformation. Knowledge of the fundamental elastic constants, rather than of specific properties, is instead found be of especial interest because they closely govern a wide spectrum of properties which range from machinability to tribology. Various simple criteria can be applied to the values of the elastic constants in order to predict possibly useful mechanical properties. The machinability, for example, is quite closely correlated with the ratio of the bulk modulus to the $C_{44}$ elastic constant, while ductility is related to the value of the so-called Pugh ratio of the bulk modulus to the shear modulus. In what follows, these considerations are explored for a wide selection of representative MAX phases, with priority being given to the most recent theoretical and experimental results complete to early 2021.

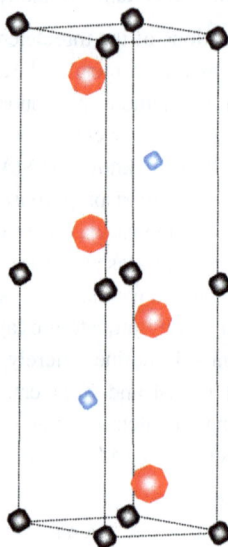

*Figure 1. Structure of Cr₂AlC*
*grey: carbon, red: chromium, blue: aluminium*
*(a = 2.86Å, c = 12.8Å, theoretical density = 5.24g/cm³)*

**Cr₂AlC**

*Hardness*

Thin (1.2μm) crystalline films of this phase (figure 1), first reported in 1963[1], were deposited[2] layer-by-layer, using magnetron sputtering of elemental targets, onto polished Inconel 718 substrates at 853K. The hardness was about 15GPa and the Young's modulus was about 260GPa. The films did not delaminate, and exhibited ductile behaviour during nanoscratching. Nanolamellar coatings with a columnar structure and nanocrystalline

Materials Research Forum LLC

https://doi.org/10.21741/9781644901274

sub-structure were direct-current magnetron sputter-deposited[3] onto Inconel 718 superalloy or (100) silicon wafers and the properties were measured by nano-indentation. The deposition rate increased with sputtering power and the coatings comprised $Cr_2AlC$ $AlCr_2$ and $Cr_7C_3$ carbide phases, while there was a change in the preferred growth orientation. The hardness ranged from 11 to 14GPa, and increased slightly with sputtering power. The mechanical properties (figures 2 and 3) of high-density pure samples at up to 980C were determined[4] by nano-indentation before and after oxidation at 1200C for over 29h. There was only a slight reduction in hardness and modulus at up to 980C; implying that there was no change in the deformation mechanism. In further work[5], the film thickness decreased from 8.95 to 6.98μm with increasing bias voltage. Coatings which were deposited at 90V exhibited the least (33nm) roughness and grain-size (76nm), combined with the greatest (15.9GPa) hardness.

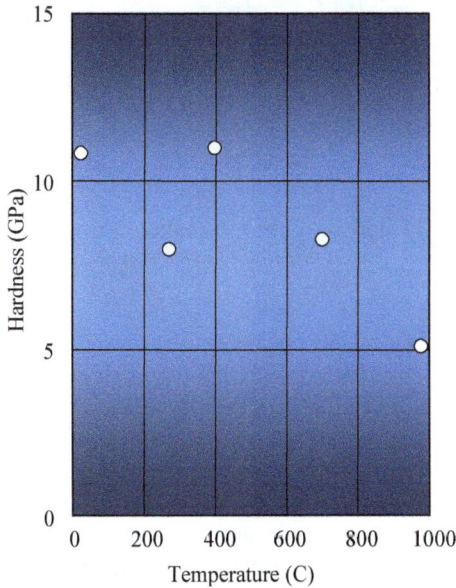

*Figure 2. Indentation hardness of $Cr_2AlC$*
*as a function of temperature*

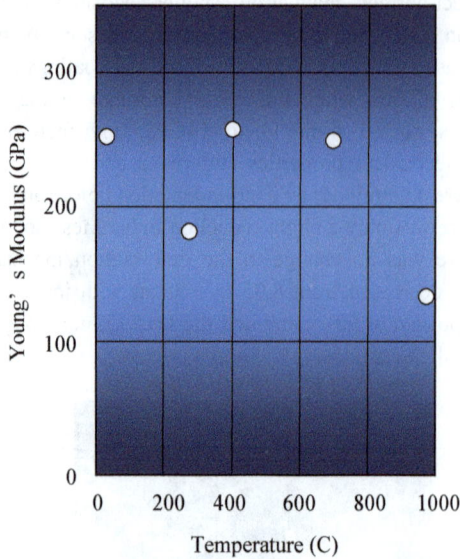

*Figure 3. Young's modulus of Cr₂AlC*
*as a function of temperature*

The effect of the aluminium content upon the mechanical properties was investigated[6] in vacuum-annealed Cr-Al-C coatings which had been deposited by the co-sputtering of $Cr_2Al$ and aluminium targets in a $CH_4/Ar$ atmosphere. Following vacuum-annealing (750C, 1.5h), the coatings comprised $Cr_2AlC$, $Al_8Cr_5$ and $Cr_7C_3$. With increasing aluminium content, the hardness and modulus of the coatings increased from 10.17 to 19.00GPa and from 198.43 to 267.62GPa, respectively. The toughness clearly decreased. The excess aluminium led to $Al_8Cr_5$ and aluminium segregation at grain boundaries and to deterioration of the mechanical properties. When $Cr_2AlC$ coatings were prepared on high-speed steel substrates by direct-current magnetron sputtering, combined with subsequent heat-treatment[7], their formation was suppressed by an increased $CH_4/Ar$ flow-rate and increased annealing temperature. This could be corrected by annealing at 750C. The nanohardness and elastic modulus of the coatings ranged from 10.6 to 17.6GPa and from 255.3 to 323.4GPa, respectively.

When $Cr_2AlC$ was irradiated[8] with 7MeV $^{26}Xe^+$ and 500keV $He^{2+}$ ions at room temperature, there was an increase in the *c* lattice parameter and a decrease in the *a* lattice parameter following irradiation to doses above 1dpa. The modified structure was stable up to a dose of 5.2dpa, without apparent lattice disorder. This suggested that there was a saturation, of irradiation effects, that was due to irradiation-induced antisite defects or carbon interstitials. In this connection, the formation energies of various defects were calculated (table 1). The interstitial position which was associated with the greatest free volume could be represented by an octahedron which was located between the chromium and aluminium layers. This was also the position which was occupied by added carbon atoms in the modified structure. Calculations were not performed for antisite defects such as $C_{Cr}$, $C_{rc}$, $C_{Al}$ and $Al_C$ because their formation energies were anticipated to be prohibitively high. The chromium and aluminium vacancies had comparable formation energies, and these were higher than that of the carbon vacancy. This suggested that carbon atoms could be more easily dislodged than could chromium and aluminium atoms. Among the interstitial defects, the carbon interstitial had the lowest formation energy. This suggested that this would be the most stable interstitial in the octahedra located between the chromium and aluminium layers. The formation energy of $Al_{Cr}$ was higher than that of $Cr_{Al}$, and both antisite defects had much lower formation energies than those of chromium or aluminium interstitials. This implied that, if displaced, the chromium and aluminium atoms would become antisite defects rather than interstitials.

*Table 1. Calculated formation energies of defects in Cr₂AlC*

| Defect | Formation Energy (eV) |
|--------|----------------------|
| $V_{Cr}$ | 1.936 |
| $V_{Al}$ | 2.090 |
| $V_C$ | 0.976 |
| $Cr_i$ | 4.526 |
| $Al_i$ | 5.226 |
| $C_i$ | 2.192 |
| $Cr_{Al}$ | 0.982 |
| $Al_{Cr}$ | 1.362 |

A simple electrochemical de-oxidation method was used[9] to produce high-quality homogeneous micron-sized particles of $Cr_2AlC$ from oxides of chromium and aluminium in a molten $NaCl$-$CaCl_2$ electrolyte. The resultant powder could be easily pressed to give a monolithic bulk product, with a hardness of 4.8GPa, which exhibited an excellent oxidation resistance at below 1100C. Single-phase bulk material was prepared[10] by *in situ* hot-pressing and solid-liquid reaction. The lattice constants were a = 2.858Å and c = 12.818Å, and the measured Vickers hardness was 5.5GPa; twice that of $Ti_2AlC$.

The monovacancy-related properties were investigated[11] by performing first-principles calculations, and these indicated that a carbon vacancy could form easily due to the low required energy. The Vickers hardness was expected to be reduced by the formation of vacancies. On the other hand, the bulk modulus was expected to be increased by the introduction of monovacancies, while they would have a negative effect upon the shear modulus and brittleness.

A chromium aluminium carbon target with a 2:1:1 molar ratio was hot-pressed at 650C to yield $Cr_2AlC$ coatings via arc ion plating at room temperature, followed by annealing[12]. The as-prepared coating had a multilayer structure which was perpendicular to the growth direction. It had a hardness of 8.8GPa. Polycrystalline coatings were prepared[13] on M38G superalloy by using a two-step method which involved magnetron sputtering from Cr-Al-C targets at room temperature, and subsequent annealing (620C). Regardless of the target composition, when the molar ratio of chromium, aluminium and carbon in the starting materials was 2:1:1, pure crystalline coatings were produced. These coatings were dense, crack-free and had a duplex structure. Following annealing (620C, 20h, argon), the hardness was 12GPa.

High-purity dense samples, prepared by hot-pressing, exhibited delamination, kink-bands, monolamellar kinking, transgranular cracking and transgranular fracture of the bulk $Cr_2AlC$ during room-temperature testing[14]. The density, Vickers hardness, flexural strength, Young's modulus, compressive strength and fracture toughness were $5.17g/cm^3$, 4.9GPa, 469MPa, 282GPa, 949MPa and $6.22MPam^{0.5}$, respectively. The strength of $Cr_2AlC$ was greatly improved by the presence of second-phase $Cr_7C_3$, but the slip of basal planes and slip systems could be hindered by this carbide, thus resulting in a lower toughness.

Dense $Cr_2AlC$ with a grain size of about 2μm could be prepared[15] at the relatively low temperature of 1100C. The fine-grained specimens had a relative density of 99%; higher than the 95% relative density of coarse-grained (35μm) $Cr_2AlC$ made by the hot-pressing of non-milled chromium, aluminium and carbon. The flexural strength and Vickers hardness of the dense fine-grained material were 513MPa and 6.4GPa, respectively.

These values were much higher than those, 305MPa and 3.5GPa respectively, for less dense and coarse-grained samples. The fracture toughness of the fine-grained material was $4.7MPam^{0.5}$, and this was lower than the value of $6.2MPam^{0.5}$ which was found for coarse material. When powder (9μm) was sintered[16] to give a relative density of about 95.5%, the sintering was characterized by an activation energy of 267kJ/mol and the mixed sintering mode involved a surface diffusion mechanism. The sintered samples had a Vickers hardness of 3.4GPa.

Bulk samples were prepared[17] by spark plasma sintering of coarse or fine powders at 1100 to 1400C. The $Cr_2AlC$ major phase always contained minor or trace amounts of $Cr_7C_3$ and $Cr_2Al$, respectively, in both fine-powder and coarse-powder derived samples, and the amounts of those two phases decreased with increasing temperature. The $Cr_2AlC$ phase content in fine-powder derived specimens was higher than that in the coarse-powder specimens when sintered at the same temperature. Grains of smaller size and fewer pores appeared in spark plasma sintered fine-powder specimens as compared with coarse-powder specimens. As a result, the hardness was higher (5.6GPa) for fine-powder material than for coarse-powder material (3.9GPa). Dense and largely single-phase samples, with perhaps a trace of $Cr_7C_3$, were prepared[18] by hot-pressing (1400C, 1h) the elemental powders. Here the hardness, Young's modulus, flexural strength and compressive strength were 5.2GPa, 288GPa, 483MPa and 1159MPa, respectively; values which were comparable to those for $Ti_3AlC_2$ and $Nb_2AlC$. The product also exhibited a good damage-tolerance. At indentation loads of up to 50N, the post-indentation flexural strength decreased by some 10%. It decreased by 31% for a load of 100N. The flexural strengths of samples which were quenched from 300C to room temperature decreased from 483 to 455MPa. The retained strength decreased to 199MPa when the quench temperature was increased to 500C. Further increases of the quench temperature to 700, 900 or 1100C resulted in a small strength reduction. When 10 to 60at% of additional aluminium was used[19] in the starting composition, the amount of $Cr_2AlC$ in samples which were hot-pressed under argon was increased and the amount of $Cr_7C_3$ second phase was decreased, such that $Cr_2AlC$ became the only phase that was present when the amount of added aluminium was more than 10at%. The bulk density gradually decreased however with increasing aluminium content. Preparation under nitrogen rather than argon resulted in the presence of AlN in the samples. As before, the hardness increased from 3.5GPa, when using coarse powder, to 4.5GPa when fine powder was used. In similar studies[20], the samples were single-phase when the starting composition contained more than 20at% of excess aluminium. The bulk density again decreased with increasing excess aluminium. The room-temperature hardness, flexural strength and Young's modulus of samples which were hot-pressed (1400C, 1h) were 3.5GPa, 375MPa and 278GPa, respectively.

*Table 2. Mechanical properties parallel to the sintering compaction direction of Cr₂AlC*

| Density (g/cm³) | Young's Modulus (GPa) | Flexural Strength (MPa) | Toughness (MPam$^{0.5}$) |
|---|---|---|---|
| 4.87 | 298 | 332 | 6.34 |
| 4.87 | 288 | 386 | 6.65 |
| 4.94 | 282 | 461 | 7.52 |
| 4.95 | 299 | 791 | 8.70 |
| 4.97 | 330 | 759 | 7.11 |

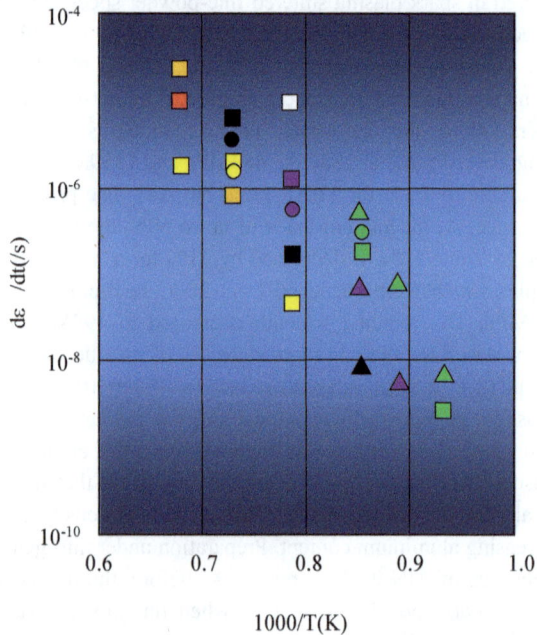

*Figure 4. Minimum creep rate of dense (2%-porosity) Cr₂AlC as a function of temperature. Squares: heating. Circles: cooling. Triangles: heating under argon. Red: 1MPa. Orange: 1.5MPa. Yellow: 2MPa. Black: 3MPa. Purple: 5MPa. Green: 10MPa. White: 12MPa.*

Specimens were prepared[21] by the reactive hot-pressing (1200 to 400C, 25MPa, argon) of CrC$_{0.5}$ and aluminium powders. Fully dense high-purity material could be produced by using hot-pressing temperatures which were as low as 1200C. The samples had a typical layered structure, with a grain-size of between 10 and 100μm; depending upon the hot-pressing temperature. The Vickers hardness (10N) of the bulk material was less than 4GPa, and this decreased with increasing indentation load. The maximum flexural strength was greater than 600MPa. The Cr$_2$AlC was easily machinable, and damage-tolerant.

Samples were prepared[22] by ball-milling and hot-pressing (1300C, 2h, 30MPa) of the resultant powder. The grain-size of the powder was uniform and smaller following the ball-milling. The flexural strength, fracture toughness and Vickers hardness were markedly increased by using powder which was ball-milled at 200rpm. The organization of hexagonal dislocation networks during hot-pressing, and the reduced grain size, improved the mechanical properties. The flexural strength and fracture toughness (table 2) reached maximum values of 791MPa and 8.70MPam$^{0.5}$, respectively.

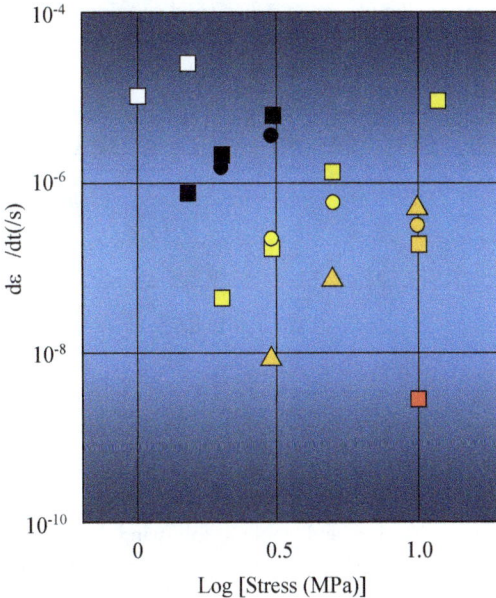

*Figure 5. Minimum creep rate of dense (2%-porosity) Cr$_2$AlC as a function of stress. Squares: heating. Circles: cooling. Triangles: heating under argon. Red: 1073K. Orange: 1173K. Yellow: 1273K. Black: 1373K. White: 1473K.*

Materials Research Forum LLC

https://doi.org/10.21741/9781644901274

When the mechanical properties of $Cr_2AlC$ cantilevers and doubly-clamped beams were investigated by using *in situ* bending tests[23], the structures exhibited a Young's modulus of 184GPa. This was close to the value which was deduced from vibrational measurements, but the bending tests exhibited a lower variance of the results.

A $Cr_2Al(Si)C$ solid solution, with traces of $Al_2O_3$ and $Cr_5Si_3$, was prepared[24] by hot-pressing (1450C, 1h, 30MPa, argon) a mixture of chromium, aluminium, silicon and carbon powders in the molar ratio of 2:1.1:0.2:1. The solid solution, with a maximum silicon content of 3.5at% and a grain size of 50μm, had a layered structure and buckling and kinking of the layered structure - as well as grain delamination crack-deflection – was extensive around indentations and at fracture surfaces. The latter energy-absorbing mechanisms imparted considerable room-temperature damage-tolerance and micro-scale plasticity. The hardness, flexural strength and fracture toughness were about 3.0GPa, 354MPa and $6.6MPam^{0.5}$, respectively. These flexural strength and fracture toughness values were higher than those (305MPa and $6.2MPam^{0.5}$) for $Cr_2AlC$. The Vickers hardness of the solid solution was a little lower than that (3.4GPa) of $Cr_2AlC$; suggesting that the Hall-Petch strengthening was greater than solid-solution hardening.

*Table 3. Parameters of compressive creep for porous $Cr_2AlC$*

| Porosity (%) | State | Activation Energy (kJ/mol) | Stress Exponent |
|---|---|---|---|
| 1.9 | heating | 429 | 2.7 |
| 53 | heating | 242 | 1.9 |
| 53 | cooling | 506 | 1.9 |
| 75 | heating | 219 | 1.7 |
| 75 | cooling | 502 | 4.8 |

*Creep*

The creep behaviours of dense (2%) and porous (53% or 75%) samples were measured[25] at 1073 to 1473K during heating and cooling following high-temperature exposure (figures 4 and 5, table 3). Compressive tests were performed using stresses ranging from 1 to 12MPa. The creep rates of porous samples were higher than those of dense samples at lower temperatures, but were lower at higher temperatures. This was attributed to the effects of oxide scales and to the crack-healing which was associated with that scale formation. It had previously been observed that MAX phases such as $Ti_3AlC_2$, $Ti_2AlC$

and Cr$_2$AlC self-healing behaviour: cracks were filled and thereby healed by oxidation products of the M and A elements of the MAX phase at high operating temperatures. Following crack-healing, the fracture strength returned to the original level. Cracks with widths of 500nm and 4μm could be healed by 3 to 5h of annealing at 900 and 1200C, respectively[26].

*Table 4. Properties of M$_2$AlC phases*

| Phase | Density (g/cm$^3$) | G (GPa) | E (GPa) | B (GPa) | Poisson Ratio |
|-------|--------------------|---------|---------|---------|---------------|
| Ti$_2$AlC | 4.1 | 118 | 277 | 144 | 0.19 |
| V$_2$AlC | 4.81 | 116 | 277 | 152 | 0.20 |
| Cr$_2$AlC | 5.24 | 105 | 245 | 138 | 0.20 |
| Nb$_2$AlC | 6.34 | 117 | 286 | 165 | 0.21 |

### Elastic constants

A systematic investigation[27], at 5 to 300K, of the elastic properties of M$_2$AlC phases, using sound-velocity and diamond-anvil cell methods, revealed Young's and shear moduli of 270GPa and 120GPa level (table 4), respectively, indicating that the phases were quite stiff.

*Table 5. Calculated elastic constants of M$_2$AlC phases*

| Phase | C$_{11}$ (GPa) | C$_{12}$ (GPa) | C$_{13}$ (GPa) | C$_{33}$ (GPa) | C$_{44}$ (GPa) |
|-------|----------------|----------------|----------------|----------------|----------------|
| Ti$_2$AlC | 321 | 76 | 100 | 318 | 144 |
| V$_2$AlC | 338 | 92 | 148 | 328 | 155 |
| Cr$_2$AlC | 396 | 117 | 156 | 382 | 173 |
| Nb$_2$AlC | 334 | 115 | 149 | 324 | 154 |
| Ta$_2$AlC | 354 | 140 | 159 | 356 | 172 |

Calculations were made[28] of the elastic properties (table 5) of M$_2$AlC, where M was titanium, vanadium, chromium, niobium or tantalum, by using first-principles total-energy methods and the projector augmented-wave approach. The Young's, bulk (Voigt or Reuss) and shear moduli, and Poisson ratio for ideal polycrystalline aggregates were estimated. The elastic modulus of Cr$_2$AlC was 357.7GPa (table 6), and the values for the other phases were of the order of 309GPa. See also under $Ti_2AlN$. The differences were attributed to the energies of M-C bonds (table 7). The materials exhibited a marked elastic anisotropy.

*Table 6. Calculated shear moduli of $M_2AlC$ phases*

| Phase | E (GPa) | $B_V$ (GPa) | $B_R$ (GPa) | $G_V$ (GPa) | $G_R$ (GPa) | Poisson Ratio |
|---|---|---|---|---|---|---|
| $Ti_2AlC$ | 304.8 | 168.0 | 163.9 | 127.8 | 127.4 | 0.194 |
| $V_2AlC$ | 307.9 | 197.8 | 192.3 | 127.7 | 121.3 | 0.236 |
| $Cr_2AlC$ | 357.7 | 225.8 | 225.7 | 146.9 | 142.6 | 0.236 |
| $Nb_2AlC$ | 298.8 | 202.0 | 208.3 | 122.2 | 115.5 | 0.257 |
| $Ta_2AlC$ | 318.6 | 220.0 | 222.2 | 130.6 | 122.2 | 0.260 |

Mechanical properties are unfortunately not the only consideration in materials use and, in spite of a very good oxidation resistance, $Cr_2AlC$ tends to fail at temperatures above 1000C because it succumbs to spallation. This in turn arises from the phase's relatively high thermal expansion coefficient.

*Table 7. Calculated bond-energies of $M_2AlC$ phases*

| Phase | Bond | Energy (eV) |
|---|---|---|
| $Ti_2AlC$ | M-C | 2.61 |
| $Ti_2AlC$ | M-Al | 0.98 |
| $V_2AlC$ | M-C | 2.77 |
| $V_2AlC$ | M-Al | 1.09 |
| $Cr_2AlC$ | M-C | 2.94 |
| $Cr_2AlC$ | M-Al | 1.18 |
| $Nb_2AlC$ | M-C | 2.80 |
| $Nb_2AlC$ | M-Al | 1.21 |
| $Ta_2AlC$ | M-C | 2.84 |
| $Ta_2AlC$ | M-Al | 1.25 |

## Cr₂GeC

### Hardness

Polycrystalline samples were prepared[29] via the hot-pressing (1350C, 6h, 45MPa) of pure chromium, germanium and carbon powders. They were easily machinable and fully dense. The Vickers hardness was 2.5GPa and the Young's modulus was 200GPa. The shear modulus and Poisson ratio, as deduced from ultrasound data, were 80GPa and 0.29, respectively. The ultimate compressive strength for specimens with a 20μm grain-size was 770MPa. Samples which were compressively loaded from 300 to 570MPa exhibited non-linear fully-reversible closed hysteresis loops which dissipated some 20% of the mechanical energy. The energy dissipation was attributed to the formation and annihilation of kink-bands. The critical resolved shear stress of basal plane dislocations was estimated to be 22MPa, while the kink-band and reversible dislocation densities at a stress of 568MPa were estimated to be $1.2 \times 10^{-2}/\mu m^3$ and $1.0 \times 10^{10}/cm^2$, respectively.

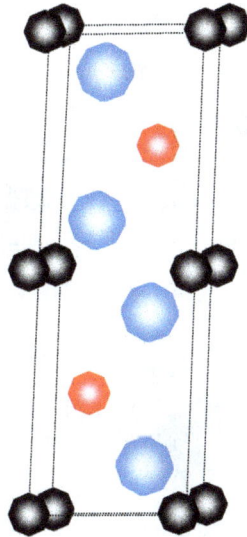

*Figure 6. Structure of Cr₂GeC*
*grey: carbon, red: germanium, blue: chromium*
*(a = 2.95Å, c = 12.08Å, theoretical density = 6.88g/cm³)*

*Table 8. Bulk moduli and pressure derivatives of M₂GeC phases*

| Phase | B (GPa) | dB/dP |
|---|---|---|
| Ti₂GeC | 160 | 4.36 |
| V₂GeC | 185 | 4.59 |
| Cr₂GeC | 214 | 4.26 |
| Zr₂GeC | 149 | 4.36 |
| Nb₂GeC | 207 | 4.35 |
| Mo₂GeC | 225 | 4.55 |
| Hf₂GeC | 167 | 4.73 |
| Ta₂GeC | 243 | 4.55 |
| W₂GeC | 244 | 4.54 |

*Table 9. Calculated elastic constants of M₂GeC phases*

| Phase | C₁₁ (GPa) | C₃₃ (GPa) | C₄₄ (GPa) | C₁₂ (GPa) | C₁₃ (GPa) | C₆₆ (GPa) |
|---|---|---|---|---|---|---|
| Ti₂GeC | 279 | 283 | 125 | 99 | 95 | 90.0 |
| V₂GeC | 282 | 259 | 160 | 121 | 144 | 80.5 |
| Cr₂GeC | 315 | 354 | 89 | 148 | 146 | 83.5 |
| Zr₂GeC | 224 | 243 | 99 | 105 | 108 | 59.5 |
| Nb₂GeC | 308 | 306 | 177 | 133 | 168 | 87.5 |
| Mo₂GeC | 331 | 342 | 123 | 136 | 184 | 97.5 |
| Hf₂GeC | 269 | 278 | 128 | 96 | 125 | 86.5 |
| Ta₂GeC | 370 | 389 | 220 | 147 | 194 | 111.5 |
| W₂GeC | 340 | 368 | 117 | 146 | 222 | 97.0 |

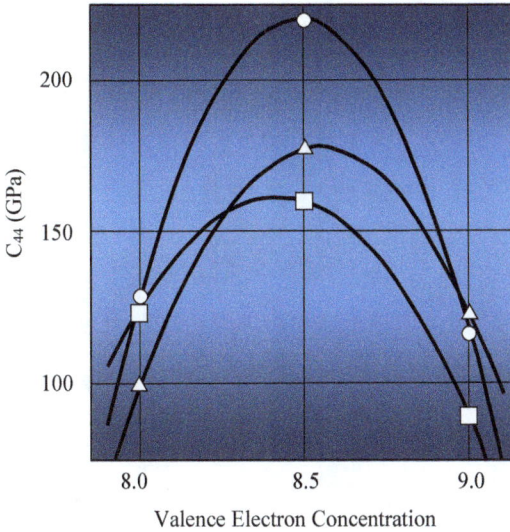

*Figure 7. Elastic constant, $C_{44}$, of $M_2GeC$ phases as a function of valence-electron concentration per atom, circles: hafnium, tantalum, tungsten, triangles: zirconium, niobium, molybdenum, squares: titanium, vanadium, chromium*

### Elastic constants

First-principles calculations were used to determine the structural (figure 6) and elastic properties (table 8) of $M_2GeC$ phases, where M was titanium, vanadium, chromium, zirconium, niobium, molybdenum, hafnium, tantalum or tungsten[30]. The elastic constants (table 9) were estimated by using the static finite strain technique. The shear modulus, $C_{44}$, which was directly related to the hardness, attained its maximum value when the valence-electron concentration was between 8.41 and 8.50 (figure 7). The Young's, bulk (Voigt, Reuss) and shear moduli and Poisson ratio (table 10) were deduced for ideal polycrystalline aggregates.

*Table 10. Calculated shear moduli of $M_2GeC$ phases*

| Phase | $B_R$ (GPa) | $B_V$ (GPa) | $G_R$ (GPa) | $G_V$ (GPa) | E (GPa) | Poisson Ratio |
|-------|-------|-------|-------|-------|-------|-------|
| $Ti_2GeC$ | 157.6 | 157.6 | 102.3 | 104.8 | 254.9 | 0.231 |
| $V_2GeC$ | 182.3 | 182.3 | 91.3 | 107.7 | 252.6 | 0.269 |
| $Cr_2GeC$ | 206.6 | 207.0 | 88.2 | 88.6 | 232.1 | 0.313 |
| $Zr_2GeC$ | 147.8 | 148.1 | 71.9 | 76.2 | 190.4 | 0.286 |
| $Nb_2GeC$ | 206.0 | 206.6 | 100.1 | 118.5 | 278.7 | 0.275 |
| $Mo_2GeC$ | 221.7 | 223.5 | 97.4 | 102.0 | 260.2 | 0.305 |
| $Hf_2GeC$ | 166.8 | 167.5 | 94.2 | 99.8 | 243.9 | 0.257 |
| $Ta_2GeC$ | 242.5 | 244.3 | 129.3 | 149.9 | 351.5 | 0.259 |
| $W_2GeC$ | 240.4 | 247.5 | 88.9 | 96.7 | 247.2 | 0.331 |

## $(Cr,Ti)_3AlC_2$

### *Hardness*

Bulk $(Cr_{0.67}Ti_{0.33})_3AlC_2$ was synthesized[31] by *in situ* reaction and hot pressing using chromium, titanium, aluminium and carbon powder. Measurement of the mechanical properties indicated that the Vickers hardness of the present material was 5.6GPa, and was thus higher than that of $Ti_3AlC_2$. The room-temperature flexural and compressive strengths were 493 and 1407MPa, respectively. The flexural strength retained a value of 424MPa, some 90% of that at room temperature, at up to 1000C. The change in the elastic modulus with increasing temperature was similar to that of the flexural strength.

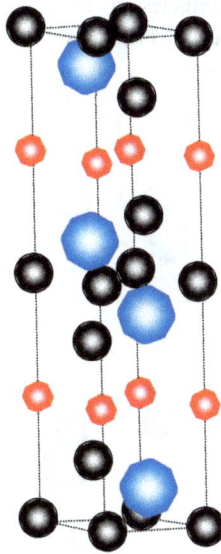

*Figure 8. Crystal structure of Hf₃AlC₂*
*black: hafnium, red: aluminium, blue: carbon*

## Hf₃AlC₂

### *Hardness*

A general study[32] of MAX $Cr_2AlC$-structured phases of the form, $Hf_2XY$, where X was aluminium, silicon or phosphorus and Y was boron, carbon and/or nitrogen, was performed using first-principles density functional plane-wave pseudopotential calculations within the generalized gradient approximation. Most of the compounds were energetically stable, and the calculated elastic constants and phonon dispersion curves showed that they were mechanically stable. The boron-containing ones, apart from $Hf_2PB$, were dynamically unstable. All of the stable compounds were metallic. The ground-state physical properties of $Hf_3AlC_2$ (figure 8) were investigated[33] using first-principles density functional theory. Calculated elastic constants and phonon dispersions confirmed the mechanical and dynamic stability of the compound. A high bulk modulus, combined with a low shear resistance and a low Vickers hardness, implied good metal-like machinability. There was nevertheless an appreciable stiffness that was due to a high Young's modulus, and a tendency to brittleness; with the Poisson ratio and Cauchy pressure being comparable to those of a ceramic. Calculations also showed that the material is elastically anisotropic.

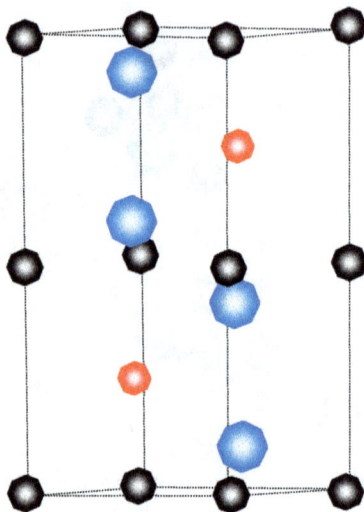

*Figure 9. Structure of Hf₂InC*
*grey: carbon, blue: hafnium, red: indium*
*(a = 3.30Å, c = 14.73Å, theoretical density = 11.57g/cm³)*

*Table 11. Properties of Hf$_2$InC as a function of pressure*

| Pressure (GPa) | Young's Modulus (GPa) | Poisson Ratio |
|:---:|:---:|:---:|
| 0 | 267.41552 | 0.20991 |
| 5 | 292.7699 | 0.22837 |
| 10 | 317.25013 | 0.23457 |
| 15 | 337.86782 | 0.24732 |
| 20 | 358.45017 | 0.25619 |
| 25 | 382.29505 | 0.26128 |
| 30 | 397.18274 | 0.27215 |
| 35 | 409.91731 | 0.27676 |
| 40 | 427.9807 | 0.28201 |
| 45 | 436.33975 | 0.29055 |
| 50 | 447.23325 | 0.29495 |

*Table 12. Pressure derivatives of the elastic constants of Hf$_2$InC*

| Derivative | Value |
|:---:|:---:|
| $\partial C_{11}/\partial P$ | 5.06 |
| $\partial C_{33}/\partial P$ | 6.06 |
| $\partial C_{44}/\partial P$ | 1.83 |
| $\partial C_{12}/\partial P$ | 3.22 |
| $\partial C_{13}/\partial P$ | 3.83 |
| $\partial B/\partial P$ | 4.21 |

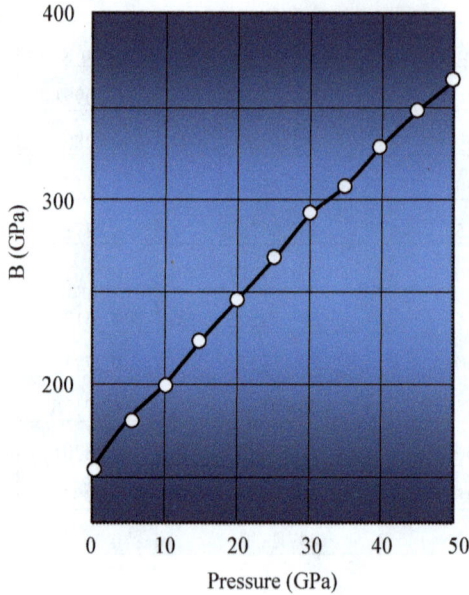

*Figure 10. Bulk modulus of Hf₂InC as a function of pressure*

## Hf$_2$InC

### Elastic constants

By using a plane-wave pseudopotential method which was based upon density functional theory within the local density approximation, a study was made[34] of the structural (figure 9) and elastic properties of this MAX phase. The effect of high pressures upon the lattice parameters was such that the contraction along the c-axis was greater than that along the a-axis. The pressure dependences of the elastic moduli (table 11, figures 10 and 11) were linear. The elastic constants, $C_{11}$, $C_{33}$ and $C_{44}$ increased with pressure, but the first two were very close and were at least twice as large as the shear elastic modulus, $C_{44}$. All of the elastic moduli, apart from $C_{44}$, had giant positive pressure derivatives (table 12).

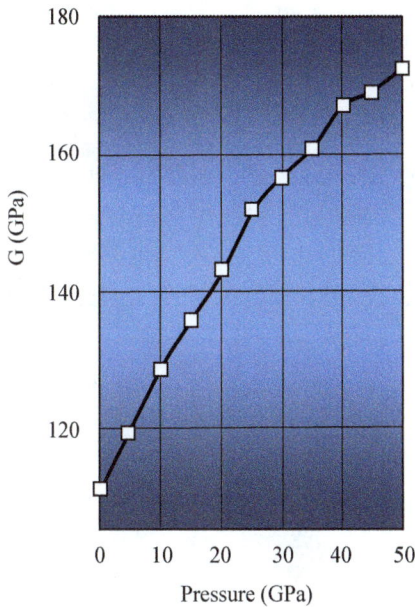

*Figure 11. Shear modulus of Hf₂InC as a function of pressure*

## Hf₂PbC

### *Elastic constants*

The effect of pressures of up to 40GPa upon the properties of Hf₂PbC was investigated[35] by using first-principles methods and density functional theory. An increase in the elastic constants with pressure (table 13) caused the material to become more resistant to shear deformation.

*Table 13. Calculated elastic constants (GPa) of Hf₂PbC as a function of pressure*

| Pressure (GPa) | $C_{11}$ | $C_{12}$ | $C_{13}$ | $C_{33}$ | $C_{44}$ | $C_{66}$ |
|---|---|---|---|---|---|---|
| 0 | 237.07 | 71.49 | 70.77 | 217.02 | 79.50 | 82.79 |
| 10 | 285.63 | 80.20 | 142.83 | 280.04 | 102.35 | 102.71 |
| 20 | 362.62 | 118.83 | 173.06 | 350.43 | 129.17 | 121.89 |
| 30 | 424.14 | 150.60 | 205.16 | 411.85 | 150.96 | 136.77 |
| 40 | 463.79 | 192.14 | 238.77 | 471.32 | 176.31 | 135.82 |

*Table 14. Calculated bulk moduli (GPa) of Hf₂PbC as a function of pressure*

| Pressure (GPa) | B | G | E | Poisson Ratio |
|---|---|---|---|---|
| 0 | 124.03 | 80.18 | 197.91 | 0.23 |
| 10 | 174.87 | 91.46 | 233.64 | 0.27 |
| 20 | 222.44 | 114.97 | 294.21 | 0.28 |
| 30 | 264.32 | 132.48 | 340.54 | 0.29 |
| 40 | 303.74 | 143.72 | 372.43 | 0.30 |

*Table 15. Melting point of Hf₂PbC as a function of pressure*

| Pressure (GPa) | Melting Point (K) |
|---|---|
| 0 | 1390.74 |
| 10 | 1630.95 |
| 20 | 1967.50 |
| 30 | 2244.19 |
| 40 | 2452.35 |

A linear change in the bulk and Young's moduli with pressure (table 14) reflected an increase in hardness with increasing pressure. The Pugh ratio between 0 and 30GPa indicated the continued occurrence of brittle behaviour, but ductile behaviour occurred at

40GPa; together with a possible phase transition. The brittleness decreased gradually with increasing pressure, while the G/B ratio decreased. There was also an increase in anisotropy with increasing pressure. The optimized lattice parameters (figure 12) were in good agreement with available theoretical and experimental data under zero pressure. The elastic constants satisfied the Born criteria for mechanical stability at the various pressures. The relatively low Poisson ratios suggested that there existed a high degree of covalent directional bonding; both covalent and ionic bonding being present. The melting point increased with pressure (table 15).

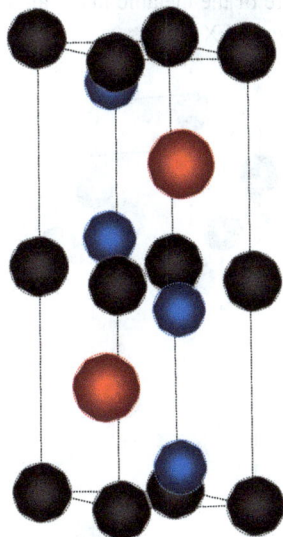

*Figure 12. Structure of Hf$_2$PbC*
*grey: carbon, blue: hafnium, red: lead*
*(a = 3.55Å, c = 14.46Å, theoretical density = 12 13g/cm$^3$)*

**Hf$_3$PB$_4$**

*Hardness*

This thermodynamically predicted MAX(MAB) phase has been studied[36] by means of first-principles calculations, giving the lattice constants (a = 3.24Å, c = 10.52Å, figure

13) of the optimized cell volume. The elastic constants (table 16) and elastic moduli (table 17) were compared with those of various MAX phases. In particular, the microhardness and macrohardness values suggested that no other MAX phase has a greater hardness, with the Vickers hardness being 7.85GPa. The remarkable elastic moduli and hardness values were explained in terms of the density-of-states and charge-density mapping. The high stiffness was attributed to the additional boron atoms, which led to strong B–B covalent bonds. A brittle nature was indicated by its Pugh ratio and Poisson ratio. A directional dependence of the elastic moduli and anisotropy factors confirmed the anisotropic nature of the metallic material. The melting point of 2282K is also higher than that of any other MAX phase.

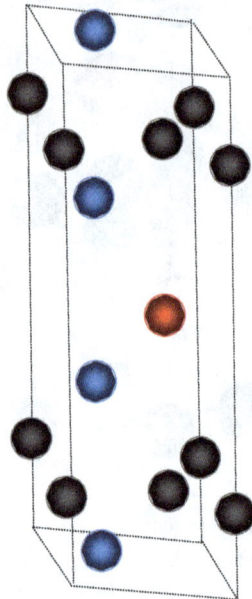

*Figure 13. Structure of Hf₃PB₄*
*grey: boron, blue: hafnium, red: phosphorus*

## Ir₂YSi

### Hardness

Two MAX-like layered materials, Rh₂YSi and Ir₂YSi, were reported[37] which had bulk moduli of 150 and 185GPa, respectively; magnitudes which were comparable to those of typical MAX phases such as Ti₂AlC, Ti₃AlC₂ and Ti₃SiC₂. On the other hand, the shear moduli were much lower than those of MAX phases, being 82 and 97GPa for Rh₂YSi and Ir₂YSi, respectively. The high stiffness was attributed to the presence of rigid Si₂–M–Si₃–M units, where M was iridium or rhodium, and the low shear deformation resistance was attributed to the presence of metallic bonds and weak bonds which linked the rigid Si₂–M–Si₃–M units. The potential slip systems for both Rh₂YSi and Ir₂YSi were {00•1}<21•0> and {11•0}<00•1>.

Table 16. Elastic constants (GPa) of Hf₃PB₄, compared with those of other phases

| Phase | $C_{11}$ | $C_{12}$ | $C_{13}$ | $C_{33}$ | $C_{44}$ |
|---|---|---|---|---|---|
| Hf₃PB₄ | 433 | 89 | 140 | 419 | 219 |
| Hf₃AlC₂ | 347 | 77 | 80 | 291 | 127 |
| Hf₃SiC₂ | 348 | 101 | 120 | 335 | 144 |
| Hf₃SnC₂ | 326 | 96 | 97 | 300 | 107 |

## Lu₂SnC

### Hardness

The elastic properties under pressure were studied[38] by means of first-principles calculations. The effect of pressures of up to 20GPa showed that the compressibility along the a-axis was higher than that along the c-axis; indicating that the Lu-C bonds were more resistant than were the Lu-Sn bonds. The bulk, Young's and shear moduli increased under pressure, implying that pressure could increase the ability to resist shape-changes and improve the stiffness and hardness. The Poisson ratio changed from 0.184 to 0.246 over the 0 to 20GPa pressure range, meaning that Lu₂SnC is brittle in nature. The elastic anisotropy increased with increasing pressure.

*Table 17. Elastic moduli (GPa) and hardness (GPa) of Hf₃PB₄, compared with other phases*

| Phase | B | G | Y | Macrohardness | Microhardness | Poisson Ratio |
|-------|-----|-----|-----|--------------|--------------|---------------|
| $Hf_3PB_4$ | 225 | 180 | 426 | 29.14 | 37.89 | 0.18 |
| $Hf_3AlC_2$ | 162 | 127 | 302 | 22.59 | 26.31 | 0.19 |
| $Hf_3SiC_2$ | 190 | 127 | 312 | 18.24 | 23.14 | 0.23 |
| $Hf_3SnC_2$ | 170 | 110 | 272 | 15.80 | 19.52 | 0.23 |

## MoAlB

### Hardness

High-purity MoAlB powder was synthesized[39] at 1100C by starting with MoB and aluminium powders in the molar ratio of 1:1.6. The excess aluminium helped to produce single-phase MoAlB, and was then removed using HCl. Bulk samples with relative densities of up to 96.7% were created by hot-pressing (1600C, 60MPa). An electrical conductivity of 2.91 x 10⁶S/m and a thermal conductivity of 29.21W/mK indicated metallic conduction behaviour. The flexural strength was 456.4MPa, the Vickers hardness was 9.3GPa, the compressive strength was 1620.2MPa and the fracture toughness was 4.3MPam^0.5. At up to 1600C, a passivating $Al_2O_3$ scale could inhibit further oxidation.

*Table 18. Calculated elastic constants of $Mo_{n+1}GeC_n$*

| Phase | $C_{11}$ (GPa) | $C_{12}$ (GPa) | $C_{13}$ (GPa) | $C_{33}$ (GPa) | $C_{55}$ (GPa) | $C_{66}$ (GPa) |
|-------|---------------|---------------|---------------|---------------|---------------|---------------|
| $Mo_2GeC$ | 295.419 | 268.572 | 242.074 | 400.010 | 29.158 | 13.424 |
| $Mo_3GeC_2$ | 426.319 | 232.663 | 235.063 | 479.669 | 139.945 | 96.828 |
| $Mo_4GeC_3$ | 416.679 | 246.219 | 255.232 | 479.891 | 115.054 | 85.230 |

## $Mo_2Ga_2C$

### *Hardness*

First-principles calculations were made[40] of this nanolaminate by comparing it with molybdenum-containing phases such as $Mo_2GaC$. The results showed that the electrical conductivity and ductility were greatly improved by adding a gallium layer, with the theoretical hardness being retained. The $Mo_2Ga_2C$ was also more metal-like. On the other hand, the extra gallium layer caused[41] a marked reduction in most of the elastic constants and moduli, the Poisson and Pugh ratios and the strength of directional bonding. Due to the extra layer, the Mo-C bond became more covalent in $Mo_2Ga_2C$ than that in $Mo_2GaC$. Finally, the extra layer reduced the hardness of $Mo_2Ga_2C$; making it relatively soft and easy to machine when compared with $Mo_2GaC$.

*Table 19. Calculated moduli of $Mo_{n+1}GeC_n$*

| Phase | E (GPa) | B (GPa) | G (GPa) | Poisson Ratio |
|---|---|---|---|---|
| $Mo_2GeC$ | 77.317 | 275.653 | 26.601 | 0.453 |
| $Mo_3GeC_2$ | 341.745 | 303.700 | 130.193 | 0.312 |
| $Mo_4GeC_3$ | 298.496 | 313.105 | 111.287 | 0.341 |

## $Mo_2GeC$

### *Elastic constants*

The structural and mechanical properties under high pressures were investigated[42] by performing first-principles calculations within the generalized gradient approximation of density functional theory. Within the pressure range of 0 to 50GPa, the material was much more compressible in the a-direction than in the c-direction. The $C_{33}$ value increased rapidly with increasing pressure, thus implying a strong resistance to compression along the c-direction. The material was predicted to be metallic, and the metallicity increased with increasing pressure. Charge-density distributions revealed the existence of covalent Mo-C bonds and, with increasing pressure, covalent Mo-Ge bonds and covalent Mo-Mo bonds.

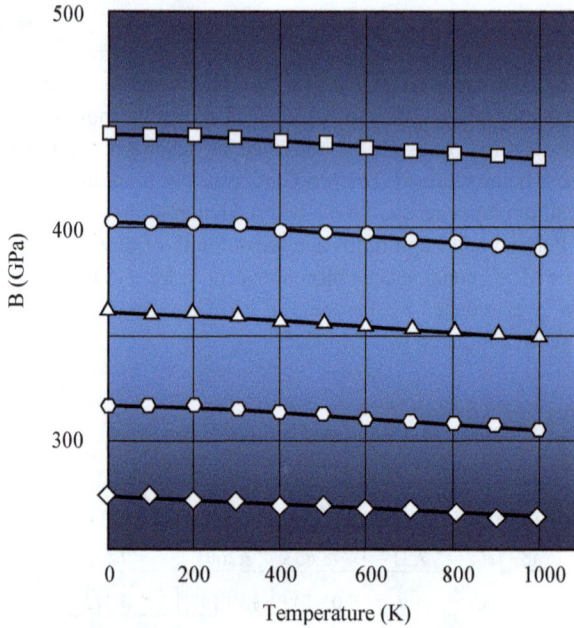

*Figure 14. Temperature dependence at various pressures of the bulk modulus of Mo₂GeC. Squares: 40GPa, circles: 30GPa, triangles: 20GPa, hexagons: 10GPa, diamonds: 0GPa*

The mechanical properties of $Mo_{n+1}GeC_n$, where n was 1, 2 or 3, were determined by using first-principles density functional theory methods[43]. These materials were metallic, and the values of the elastic constants were all positive (table 18). The elastic moduli were also calculated (table 19, figures 14 to 16). Given that B/G was greater than 1.75 and G/B was less than 0.5 for these phases, they were predicted to be ductile.

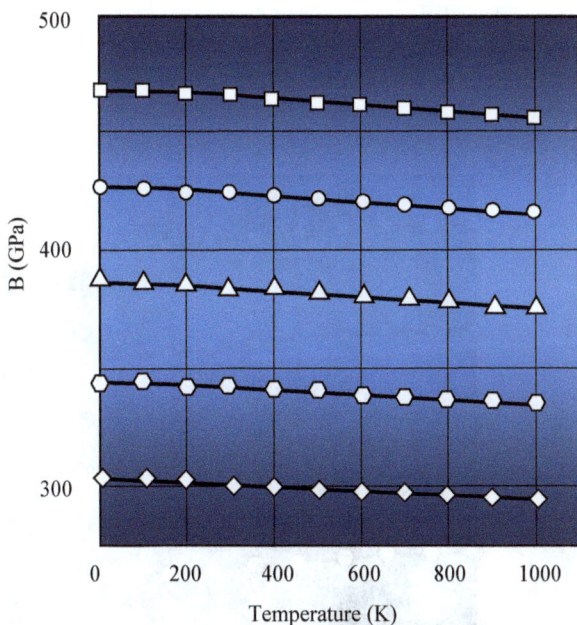

*Figure 15. Temperature dependence of the bulk modulus of $Mo_3GeC_2$. Squares: 40GPa, circles: 30GPa, triangles: 20GPa, hexagons: 10GPa, diamonds: 0GPa*

## Mo₂HoC

### *Hardness*

First-principles calculations have been used[44] to investigate the hardness and bonding strength of phases of the form: $(Mo_{0.67}R_{0.33})_2AlC$, where R is a rare earth. The Voigt-Reuss-Hill bulk, shear and Young's moduli were compared for compounds where R was neodymium, samarium, gadolinium, terbium, dysprosium, holmium, erbium, thulium or lutetium. The overall trend in properties depended upon the unit-cell volume. Nano-indentation testing of holmium-based single crystals revealed moduli which were within 10% of the predicted value, and a hardness of about 10GPa. The bonding of molybdenum and rare-earth atoms with aluminium atoms was weaker than that with carbon atoms.

This implied that exfoliation of the materials to give 2-dimensional analogues was possible, and calculated exfoliation energies indicated that this would become easier with decreasing atomic mass of the rare earth.

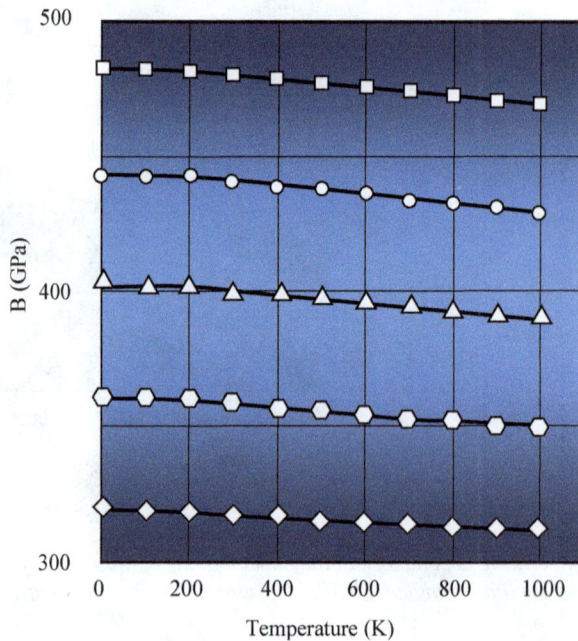

*Figure 16. Temperature dependence of the bulk modulus of $Mo_4GeC_3$. Squares: 40GPa, circles: 30GPa, triangles: 20GPa, hexagons: 10GPa, diamonds: 0GPa*

## Mo₂ScAlC₂

### *Hardness*

Density functional theory calculations were used[45] to investigate the mechanical behaviour of this material (figure 17), which was expected to exhibit a tendency to shear along the b- and c-axes when a force was applied to the a-axis. Compression along the <001>-direction under uniaxial stress was expected to be easier in this material. Volume deformation was also expected to be easier in the present material than in isostructural $Mo_2TiAlC_2$. The present material was predicted to be brittle, and cross-slip pinning was expected to be much easier in $Mo_2ScAlC_2$ than in $Mo_2TiAlC_2$. The $Mo_2ScAlC_2$ exhibited a mixture of strong covalent and metallic bonding, with a limited ionic nature. The Mo–C and Mo–Al bonds were expected to be more covalent in $Mo_2ScAlC_2$ than in $Mo_2TiAlC_2$, with the level of covalency of the Sc–C bond being relatively low when compared with that of the analogous bond in $Mo_2ScAlC_2$. Due to the lower hardness of $Mo_2ScAlC2$, it was expected to be softer and more easily machinable than $Mo_2TiAlC_2$.

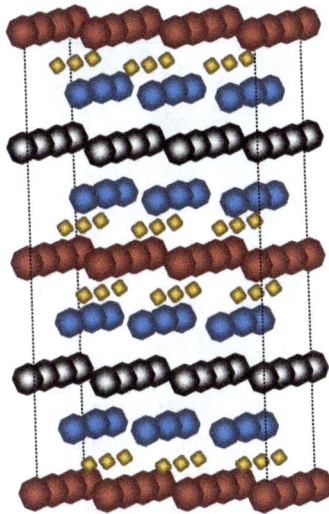

*Figure 17. Structure of chemically-ordered $Mo_2ScAlC_2$*
*red: scandium, yellow: carbon, blue: molybdenum, grey: aluminium*

## $Mo_2Ti_2AlC_3$

### *Hardness*

High-purity material was prepared by heating molybdenum, titanium, aluminium and carbon powders, and dense bulk specimens of 99% relative density were created[46] by hot-pressing (1400C, 1h, 40MPa). They exhibited the typical layered structure, with Vickers hardness of 4.81GPa, an elastic modulus of 374.15GPa, a flexural strength of 452MPa, a fracture toughness of 8.4MPam$^{0.5}$ and a compressive strength of 1145MPa (table 20). The electrical conductivity at 300 to 600K was of metallic type and decreased from 0.41 x 10$^6$ to 0.38 x 10$^6$/$\Omega$m. The thermal conductivity at 300 to 1273K decreased from 6.82 to 6.05W/mK and the thermal expansion coefficient between 350 and 1100K was 11.3 x 10$^{-6}$/K.

*Table 20. Comparison of the properties of Mo$_2$Ti$_2$AlC$_3$ and its analogues*

| Property | Mo$_2$Ti$_2$AlC$_3$ | Mo$_2$Ti$_2$AlC$_3$ | Ti$_3$AlC$_2$ | Ta$_4$AlC$_3$ |
|---|---|---|---|---|
| Density (g/cm$^3$) | 6.15 | 5.21 | 4.21 | 13.18 |
| Elastic modulus (GPa) | 374.15 | 367 | 297 | 324 |
| Vickers hardness (GPa) | 4.81 | - | 3.5 | 5.1 |
| Flexural strength (MPa) | 452 | - | 340 | 372 |
| Fracture toughness (MPam$^{0.5}$) | 8.4 | - | 7.2 | 7.7 |
| Compressive strength (MPa) | 1145 | - | 764 | 821 |

## Mo$_2$TiAlC$_2$

### Hardness

First-principles density functional theory calculations were used[47] to predict the structural and elastic properties, and the predictions were in good agreement with experimental results. Mechanical stability was confirmed by the calculated elastic constants, while brittleness was implied by the Poisson and Pugh ratios. The phase could resist compression and tension, and there was a directional bonding between the atoms. It also exhibited an appreciable elastic anisotropy, and the bonding was a mixture of covalent and metallic; plus some ionic nature. A strong directional Mo-C-Mo covalent bonding was associated with a relatively weak Ti-C bond. The predicted hardness of 9.01GPa reflected the strongly covalent bonding.

## Nb$_2$AlC

### Hardness

Polycrystalline fully-dense single-phase samples with an average grain-size of about 14$\mu$m were prepared[48] by the reactive hot isostatic pressing (1600C, 8h, 100MPa) of niobium, graphite and $Al_4C_3$. The same methods produced largely single-phase $(Ti,Nb)_2AlC$ samples with an average grain-size of $45\mu m$. In order to obtain finer-grained $(15\mu m)$ samples of the solid solution, the powder mixtures were hot-pressed at 1450C for 24h. The a- and c- lattice constants of aluminium-poor $Nb_2AlC$ samples were 3.107 and 13.888Å, respectively. The corresponding values for the solid solution were 3.077 and 13.790Å. The hardness, 5.8GPa, of the solid solution was intermediate between that of $Nb_2AlC$ and $Ti_2AlC$, but it was concluded that no solid-solution strengthening occurred in this system. All of the samples were quite damage-tolerant and thermal-shock resistant. Vickers (300N) indentation of a 1.5mm-thick bar decreased the strength by between 25 and 50%, depending upon the grain size. Quenching into water from 1200C reduced the 4-point flexural strength by 40 to 70%. The strengths of $(Ti,Nb)_2AlC$ samples decreased anomalously with increasing strain-rate.

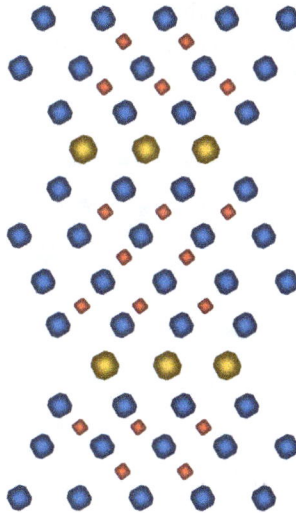

*Figure 18. Projection of the $Nb_{12}Al_3C_8$ structure along [10•0]*
*blue: niobium, yellow: aluminium, red: carbon*

### Nb$_4$AlC$_3$

#### Hardness

Samples of tungsten-doped material, $(Nb_{1-x}W_x)_4AlC_3$ where x ranged from 0 to 0.0375, were prepared[49] by *in situ* reactive hot-pressing of niobium, aluminium, graphite and tungsten powders. It was found that $Nb_{1-x}W_xC$ was dispersed in a tungsten-doped $Nb_4AlC_3$ matrix. The mechanical properties of the $Nb_4AlC_3$ were greatly improved by the doping, such that $(Nb_{0.975}W_{0.025})_4AlC_3$ exhibited a flexural strength of 483MPa, a fracture toughness of 8.5MPam$^{0.5}$ and a Young's modulus of 382GPa at room temperature. These were increased by 59, 15 and 30%, respectively, with regard to those of the original $Nb_4AlC_3$. The Vickers hardness of $(Nb_{0.9625}W_{0.0375})_4AlC_3$ was 4.8GPa, and this was 92% higher than that of $Nb_4AlC_3$. The $(Nb_{0.975}W_{0.025})_4AlC_3$ retained a flexural strength of 344MPa at 1400C, and this was equal to 71% of the room-temperature value which was, in turn, much higher than the 303MPa which was the room-temperature flexural strength of $Nb_4AlC_3$. The strengthening was attributed to the solid-solution effect of tungsten and to the presence of the $Nb_{1-x}W_xC$ second phase. When fully-dense pure $(Nb_{1-x}Ti_x)_4AlC_3$ samples, where x ranged from 0 to 0.3,

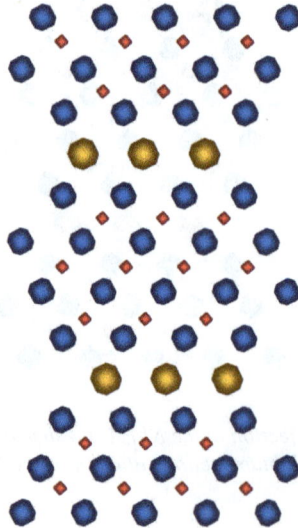

*Figure 19. Projection of the Nb$_4$AlC$_3$ structure along [11●0]
blue: niobium, yellow: aluminium, red: carbon*

*Table 21. Elastic constants of $Nb_{12}Al_3C_8$ and $Nb_4AlC_3$*

| Phase | $C_{11}$ (GPa) | $C_{12}$ (GPa) | $C_{13}$ (GPa) | $C_{33}$ (GPa) | $C_{44}$ (GPa) |
|---|---|---|---|---|---|
| $Nb_{12}Al_3C_8$ | 372.9 | 109.9 | 135.7 | 329.3 | 158.1 |
| $Nb_4AlC_3$ | 409.3 | 124.6 | 145.8 | 303.9 | 158.5 |

were prepared[50] by *in situ* reaction hot-pressing the solid solution of titanium imparted a marked strengthening to $Nb_4AlC_3$ such that $(Nb_{0.8}Ti_{0.2})_4AlC_3$ exhibited a flexural strength of 508MPa and a fracture toughness of $8.4MPam^{0.5}$. The 4.42GPa Vickers hardness of $(Nb_{0.7}Ti_{0.3})_4AlC_3)$ was 69% higher than that of $Nb_4AlC_3$. The $(Nb_{0.8}Ti_{0.2})_4AlC_3$ retained a Young's modulus of 294GPa at 1500C, and this was equal to 84% of the room-temperature value. Highly-textured polycrystalline samples were prepared[51] by slip-casting within a strong magnetic field before spark plasma sintering. The material had a layered microstructure at scales ranging from nanometers to millimetres and exhibited a greater hardness (7.0GPa), bending strength (881MPa) and fracture toughness $(14.1MPam^{0.5})$ along the c-axis. The Young's modulus was equal to 365GPa in the a- and b-axis directions. Dense bulk $Nb_4AlC_3$ was formed[52] by reaction between $Nb_2AlC$ and NbC, and the decomposition of $Nb_2AlC$. The Vickers hardness and flexural strength were 3.7GPa and 455MPa, respectively, and micro-indentation revealed an anisotropic response of the grains which reflected the anisotropic crystal structure. The flexural strength retained a value of 230MPa at up to 1400C.

*Table 22. Elastic moduli (Reuss and Voigt) of $Nb_{12}Al_3C_8$ and $Nb_4AlC_3$*

| Phase | $B_V$ (GPa) | $B_R$ (GPa) | $G_V$ (GPa) | $G_R$ (GPa) | $E_V$ (GPa) | $E_R$ (GPa) |
|---|---|---|---|---|---|---|
| $Nb_{12}Al_3C_8$ | 204.2 | 204.1 | 135.8 | 131.9 | 333.4 | 325.6 |
| $Nb_4AlC_3$ | 217.2 | 214.4 | 139.0 | 133.1 | 343.6 | 330.8 |

### Elastic constants

The carbon vacancies in $Nb_4AlC_{3-\delta}$ were recalled[53] to be ordered and to constitute an ordered phase of the form, $Nb_{12}Al_3C_8$. The spatial distribution of the ordered vacancies was clarified by comparing the $Nb_{12}Al_3C_8$ (figure 18) with vacancy-free $Nb_4AlC_3$ (figure

19) with regard to their bonding characteristics and elastic properties. The carbon vacancies were found to break only the relatively weak Nb–C bonds and to form ordered equilateral triangular carbon-vacancy networks which maximized the bond-strengthening effect. These networks tended to shift the partial and total densities of states towards the Fermi energy level and to provoke so-called de-metallization. The presence of the ordered equilateral triangular carbon-vacancy networks led to a softening of the elastic moduli (table 21) which decreased the anisotropy of the Young's modulus (table 22) while increasing that of the shear modulus.

## Nb$_2$CuC

### Hardness

A systematic density functional theory study[54], using the generalized gradient and local density approximations, has been made of the structural (figure 20) and elastic properties. The replacement of aluminium by copper modified the band profiles and thus led to improved physical properties, although Nb$_2$CuC was structurally less stable than Nb$_2$AlC. The Nb-C and Nb–A bonds (where A is Cu or Al) were covalent, and the Nb–Nb bonds led to antibonding states in both phases. Charge transfer between the constituent atoms indicated the presence of some degree of ionic character in the chemical bonds of both materials, which were mechanically and dynamically stable. It was noted that Nb$_2$CuC was ductile, relatively soft, machinable and damage-tolerant while Nb$_2$AlC was brittle. The Nb$_2$CuC was more elastically anisotropic than was Nb$_2$AlC. Samples of Ti$_2$Al$_x$Cu$_{1-x}$N and Nb$_2$CuC were synthesized[55] by reacting Ti$_2$AlN and Nb$_2$AlC, respectively, with molten CuCl$_2$ or CuI. There was complete A-site replacement in Nb$_2$AlC, leading to the formation of Nb$_2$CuC. The replacement of aluminium in Ti$_2$AlN was close to complete only for Ti$_2$Al$_{0.1}$Cu$_{0.9}$N. Density-functional theory calculations confirmed the structural stability of Nb$_2$CuC and Ti$_2$CuN, and the calculated cleavage energies in these copper-containing phases were weaker than those in the aluminium-containing equivalents.

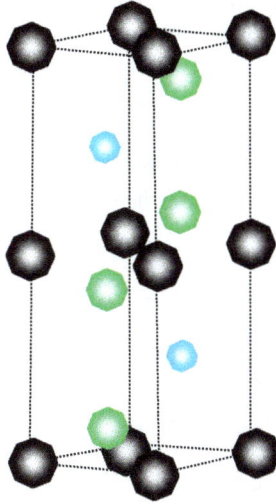

*Figure 20. Structure of Nb₂CuC*
*grey: carbon, blue: copper, green: niobium*

## Nb₂GeC

### Hardness

The monovacancy formation energies of each constituent element were deduced by performing first-principles calculations. It was found[56] that the vacancy of each element had a slight effect upon the hybridization bonding. If there was a vacancy of germanium or carbon, the hybridization bonding was slightly reduced but, if there were a niobium vacancy, the hybridization bonding was increased between the d-orbital and p-orbital. If there was a vacancy of germanium or carbon, the hardness of Nb₂GeC decreased. It increased if there was a niobium vacancy. First-principles density functional theory calculations were made[57] of the properties of the nanolaminate at 0 to 1200K. The theoretical Vickers hardness was 4.15GPa.

### Elastic constants

First-principles density functional theory methods were used[58] to calculate structural parameters and elastic constants (table 23) for Nb₂GeC. Comparison with isostructural iso-electronic materials such as $Nb_2SnC$, $V_2SnC$ and $Ta_2GeC$ suggested that the steric effect on the X-sites of 211-type MAX phases should be greater than that on the M-sites. This was in turn related to a relatively weak bonding between X atomic sheets and [$M_2C$] carbide blocks, as compared with the strong bonding within the carbide blocks. The elastic parameters indicated that $Nb_2GeC$ was mechanically stable and that the parameter which limited such stability was the shear modulus. The phase was predicted to be ductile, and to exhibit elastic anisotropy. Band-structure calculations indicated that the phase is metal-like. The anisotropic intra-atomic bonding was of mixed type in that, within the [$Nb_2C$] blocks, covalent-ionic Nb–C bonds and metal-like Nb–Nb bonds occurred while bonding between the adjacent [$Nb_2C$] blocks and germanium atomic sheets, involved ionic bonds that arose from [$Nb_2C$]-Ge charge-transfer together with relatively weak covalent Nb–Ge bonds. The bulk, shear and Young's moduli were intermediate between the predicted maximum and minimum values. The bulk modulus was higher than that of $Ti_2AlC$, while the shear modulus was smaller. The present material was expected to be ductile on the basis of the modulus ratios. It was proposed that these phases could be divided into two groups. In one group, the bulk modulus of the 211-type MAX phase was comparable to that of the corresponding binary carbide. In the other group, this modulus was much smaller than that of the carbide. In the present material, the calculated value of B for $Nb_2GeC$ was 234GPa while the bulk modulus of NbC was 293 to 328GPa. Therefore $Nb_2GeC$, and iso-electronic $Nb_2SnC$, were to be included in the first group; with strong coupling between the blocks and A-sheets.

*Table 23. Calculated elastic constants of $Nb_2GeC$*

| Constant | Value (GPa) |
|----------|-------------|
| $C_{11}$ | 282.7 |
| $C_{12}$ | 116.7 |
| $C_{13}$ | 167.1 |
| $C_{33}$ | 181.7 |
| B | 233.8 |
| G | 87.5 |
| E | 233.4 |

*Table 24. Calculated single-crystal elastic constants of $M_2PC$-type phases*

| Phase | $C_{11}$ (GPa) | $C_{33}$ (GPa) | $C_{44}$ (GPa) | $C_{12}$ (GPa) | $C_{13}$ (GPa) |
|---|---|---|---|---|---|
| $V_2PC$ | 363 | 385 | 206 | 115 | 163 |
| $Nb_2PC$ | 371 | 403 | 197 | 123 | 162 |
| $Ta_2PC$ | 366 | 413 | 199 | 133 | 171 |

## $Nb_2PC$

### Elastic constants

Pseudo-potential plane-wave methods, based upon density functional theory in the generalized gradient approximation, were used[59] to calculate the structural and elastic properties (tables 24 and 25) of $M_2PC$ phases, where M was vanadium, niobium or tantalum. The effects of pressures of up to 20GPa were investigated, and the elastic constants together with their pressure dependences were predicted by using the static finite strain technique. Numerical estimates were made of the moduli and Poisson ratios of ideal polycrystalline aggregates within the framework of the Voigt-Reuss-Hill approximation.

*Table 25. Calculated moduli of polycrystalline $M_2PC$-type phases*

| Phase | B (GPa) | G (GPa) | E (GPa) | Poisson Ratio |
|---|---|---|---|---|
| $V_2PC$ | 220.5 | 145.4 | 357.6 | 0.2297 |
| $Nb_2PC$ | 225.7 | 145.4 | 359.0 | 0.2348 |
| $Ta_2PC$ | 231.6 | 142.0 | 353.8 | 0.2454 |

The pressure dependence of the relative lattice parameters ($a/a_0$, $c/c_0$) fitted a quadratic relationship, with the c-direction being more resistant to contraction than was the a-direction. This suggested that the bonding in the c-direction was stiffer than that in the a-direction. The bonding of these materials was suggested to result from a mixture of covalent-ionic and metallic behaviours. All of the phases were mechanically stable, and there was a linear dependence of the bulk modulus and elastic constants upon the applied pressure. The calculated B/G ratios indicated that these phases should be ductile.

## Nb$_4$SiC$_3$

### Hardness

This layered compound was originally predicted[60] to exist by using first-principles methods. It was expected to be a metal with a covalent nature. It has a theoretical hardness of 10.86GPa. This is much higher than that of Nb$_4$AlC$_3$, but it is also more ductile. The high bulk modulus and hardness can be attributed to the strong covalent bonding. The elasticity is slightly anisotropic.

*Table 26. Properties of Nb$_2$SnC and similar compounds*

| Compound | Melting Point (C) | Elastic modulus (GPa) | Vickers Hardness (GPa) |
|----------|-------------------|------------------------|-------------------------|
| Ti$_2$SnC | 1250 | - | 3.5 |
| Zr$_2$SnC | 1275 | 178 | 3.9 |
| Nb$_2$SnC | 1390 | 216 | 3.8 |
| Hf$_2$SnC | 1335 | 237 | 3.8 |
| Zr$_2$PbC | <1300 | - | 3.2 |
| Hf$_2$PbC | <1300 | - | 3.8 |
| Ti$_3$SiC$_2$ | 2200 | 330 | 4 |

## Nb$_2$SnC

### Hardness

The properties of the nanolaminate material were deduced[61] by performing first-principles density functional theory calculations. The bulk modulus and thermal expansion coefficient were obtained by using the quasi-harmonic Debye model at 1200K. The theoretical Vickers hardness was 4.15GPa. Largely single-phase (92 to 94vol%) fully-dense samples were prepared[62] by reactively hot isostatically pressing (1200 to 1325C, 4 to 48h) stoichiometric mixtures of the component elemental powders. The Vickers hardness ranged from 3 to 4GPa[63] (table 26) and the compound was easily machinable. The Young's modulus was equal to 216GPa. The material dissociated into the transition-metal carbide and tin when heated at 1250 to 1390C.

*Table 27. Elastic constants of ScAN$_2$ phases as a function of pressure*

| Phase | Pressure (GPa) | C$_{11}$ (GPa) | C$_{12}$ (GPa) | C$_{13}$ (GPa) | C$_{33}$ (GPa) | C$_{44}$ (GPa) |
|-------|----------------|----------------|----------------|----------------|----------------|----------------|
| ScTaN$_2$ | 0 | 551 | 158 | 143 | 552 | 196 |
| ScTaN$_2$ | 30 | 703 | 225 | 229 | 686 | 239 |
| ScTaN$_2$ | 50 | 787 | 272 | 283 | 762 | 258 |
| ScTaN$_2$ | 100 | 962 | 387 | 413 | 933 | 287 |
| ScTaN$_2$ | 150 | 1102 | 500 | 536 | 1077 | 300 |
| ScNbN$_2$ | 0 | 522 | 152 | 130 | 546 | 185 |
| ScNbN$_2$ | 30 | 663 | 217 | 212 | 676 | 223 |
| ScNbN$_2$ | 50 | 740 | 261 | 264 | 752 | 240 |
| ScNbN$_2$ | 100 | 904 | 369 | 386 | 916 | 268 |
| ScNbN$_2$ | 150 | 1035 | 474 | 502 | 1057 | 280 |
| ScVN$_2$ | 0 | 580 | 145 | 121 | 167 | 572 |
| ScVN$_2$ | 30 | 624 | 213 | 204 | 206 | 728 |
| ScVN$_2$ | 50 | 698 | 257 | 257 | 220 | 812 |
| ScVN$_2$ | 100 | 849 | 369 | 384 | 239 | 994 |
| ScVN$_2$ | 150 | 972 | 477 | 504 | 247 | 1153 |

## (Nb,Zr)$_4$AlC$_3$

### *Hardness*

The solubility of zirconium in Nb$_4$AlC$_3$ was investigated[64] by combining the synthesis of (Nb$_x$Zr$_{1-x}$)$_4$AlC$_3$ solid solutions with density functional theory calculations. High-purity solid solutions were prepared by the reactive hot-pressing of NbH$_{0.89}$, ZrH$_2$, aluminium and carbon powders. A limited solubility of zirconium in the host lattice was consistent

Materials Research Forum LLC
https://doi.org/10.21741/9781644901274

with a calculated minimum in the energy of mixing. The room-temperature hardness, Young's modulus and fracture toughness, plus the high-temperature flexural strength and E value of $(Nb_{0.85}Zr_{0.15})_4AlC_3$ were investigated and compared with those of pure $Nb_4AlC_3$. There was an appreciable increase in fracture toughness: from 6.6 for pure $Nb_4AlC_3$ to $10.1 MPam^{0.5}$ for $(Nb_{0.85}Zr_{0.15})_4AlC_3$.

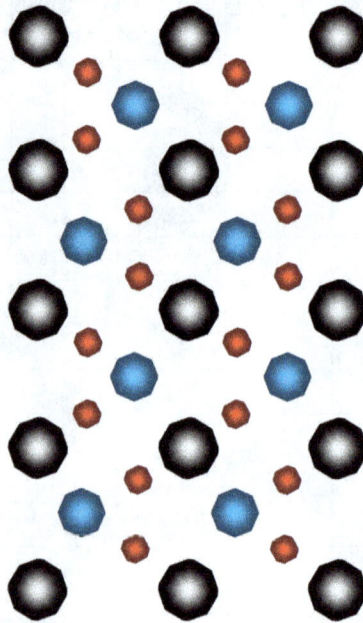

*Figure 21. Structure of ScTaN₂ as viewed along the [110] zone axis.*
*grey: scandium, blue: tantalum, red: nitrogen*

**Sc₂InB**

*Hardness*

First-principles density functional theory investigations were made[65] of $Sc_3InX$, where X was boron, carbon or nitrogen. All of the materials were brittle. The calculated Peierls stress was 3 to 5 times larger than those of various MAX phases, suggesting that dislocation movement might be much reduced. There was a stronger covalency between scandium and X atoms than that for Sc-Sc bonding. The Vickers hardness values of these materials were predicted to be between 3.03 and 3.88GPa.

*Table 28. Elastic constants (Reuss, Voigt) of ScAN₂ phases as a function of pressure (GPa)*

| Phase | Pressure | B$_R$ (GPa) | B$_V$ (GPa) | G$_R$ (GPa) | G$_V$ (GPa) | E$_R$ (GPa) | E$_V$ (GPa) |
|-------|----------|-------------|-------------|-------------|-------------|-------------|-------------|
| ScTaN₂ | 0 | 283 | 283 | 197 | 197 | 479 | 479 |
| ScTaN₂ | 30 | 384 | 384 | 244 | 245 | 604 | 606 |
| ScTaN₂ | 50 | 446 | 446 | 266 | 268 | 665 | 670 |
| ScTaN₂ | 100 | 587 | 587 | 302 | 313 | 774 | 797 |
| ScTaN₂ | 150 | 714 | 714 | 322 | 344 | 840 | 889 |
| ScNbN₂ | 0 | 268 | 268 | 189 | 190 | 460 | 460 |
| ScNbN₂ | 30 | 365 | 365 | 232 | 233 | 574 | 576 |
| ScNbN₂ | 50 | 423 | 423 | 252 | 255 | 631 | 638 |
| ScNbN₂ | 100 | 556 | 556 | 287 | 298 | 735 | 758 |
| ScNbN₂ | 150 | 676 | 676 | 306 | 327 | 798 | 844 |
| ScVN₂ | 0 | 255 | 256 | 178 | 179 | 432 | 436 |
| ScVN₂ | 30 | 356 | 357 | 222 | 224 | 551 | 556 |
| ScVN₂ | 50 | 415 | 416 | 241 | 246 | 606 | 616 |
| ScVN₂ | 100 | 548 | 551 | 271 | 285 | 697 | 729 |
| ScVN₂ | 150 | 669 | 674 | 286 | 313 | 750 | 813 |

## Sc₂InC

### *Hardness*

First-principles pseudopotential calculations were used[66] to predict the Vickers hardness. Analysis of the elastic constants, the phonon dispersion and the phonon density-of-states confirmed mechanical and dynamic stability of the material. A Mulliken population analysis reflected the prominence of covalency in the chemical bonding, and electronic charge-density mapping revealed a combination of ionic, covalent and metallic bonding. The phase was expected to be soft, and easily machinable, due to its low Vickers hardness.

## $Sc_3SnB$

### *Hardness*

The properties of $Sc_3SnX$, where X was boron or carbon, were calculated[67] using first-principles density functional theory. The mechanical stability of the materials was theoretically confirmed by using the Born criteria, and both materials were shown to be brittle. The electronic band structures revealed metallic characteristics, with the contribution arising mainly from the scandium 3d orbital. The calculated Peierls stresses suggested that dislocation movement might be much slower when compared with that in other MAX phases. The calculated Vickers hardness values of $Sc_3SnB$ and $Sc_3SnC$ were 4.45 and 4.04GPa, respectively.

## $ScTaN_2$

### *Hardness*

First-principles density functional theory calculations were used[68] to study the elastic properties at high pressures of layered $ScAN_2$ phases (figure 21), where A was vanadium, niobium or tantalum. The results showed that, as the pressure was increased from 0 to 150GPa, the elastic constants (table 27) and moduli (table 28) increased by between 53 and 216% for $ScTaN_2$, by between 72 and 286% for $ScNbN_2$ and between 61 and 317% for $ScVN_2$.

*Table 29. Elastic constants of $M_2InC$ phases*

| Phase | $C_{11}$ (GPa) | $C_{12}$ (GPa) | $C_{13}$ (GPa) | $C_{33}$ (GPa) | $C_{44}$ (GPa) |
|-------|------|------|------|------|------|
| $Zr_2InC$ | 279 | 66 | 75 | 255 | 94 |
| $Hf_2InC$ | 331 | 87 | 90 | 284 | 101 |
| $Ta_2InC$ | 452 | 147 | 197 | 425 | 161 |

## Ta₄AlC₃

### Hardness

Pressure-assisted densification was carried out[69] by hot-pressing and spark plasma sintering of $Ta_2H$, aluminium and carbon powder mixtures at 1200 to 1650C. High-purity samples of α-$Ta_4AlC_3$ were obtained by hot-pressing (1500C, 0.5h, 30MPa), while β-$Ta_4AlC_3$ was observed in samples which were produced by spark plasma sintering. The Young's modulus, Vickers hardness, flexural strength and single-edge V-notch beam fracture toughness of the high-purity bulk material were determined. Thermal decomposition into $TaC_x$ and aluminium vapour occurred under high vacuum at 1200 and 1250C.

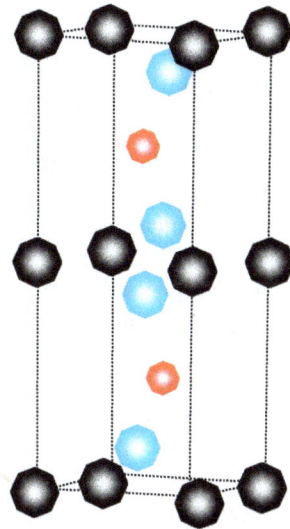

| Phase | a (Å) | c (Å) |
|-------|-------|-------|
| Zr₂InC | 3.36 | 15.11 |
| Hf₂InC | 3.36 | 14.88 |
| Ta₂InC | 3.21 | 14.57 |

*Figure 22. Structure of M₂InC phases.*
*grey: carbon, red: indium, blue: zirconium, hafnium or tantalum*

## Ta$_2$InC

### *Elastic constants*

The properties of M$_2$InC, where M was zirconium, hafnium or tantalum (figure 22), were calculated[70] using density functional theory methods. The estimated monocrystalline elastic constants (table 29) confirmed the mechanical stability of all of these compounds. The monocrystalline and polycrystalline elastic constants (table 30) increased with the atomic number of the M-element. The Pugh and Poisson ratios predicted the brittleness of the compounds, and this was associated with strong directional covalent bonding plus ionic contributions. An overlapping of the conduction and valence bands at the Fermi level explained the metallic nature of these phases. Low values of the hardness (table 31) reflected a softness of the materials, and implied easy machinability.

## Ti$_2$AlC

### *Hardness*

The microstructures of as-synthesized material were mainly single-phase and were free from amorphous grain-boundary phases[71]. High-resolution imaging revealed that the stacking sequence of titanium and aluminium atoms along the [00•1] Ti$_2$AlC direction was ABABAB. Intergrown structures, Ti$_3$AlC$_2$-Ti$_2$AlC and Ti$_2$AlC-TiC-Ti$_2$AlC, were detected by means of high-resolution imaging and energy dispersive X-ray analysis. The Ti$_3$AlC$_2$ and TiC shared close crystallographic relationships with Ti$_2$AlC, which suggests the possibility of tailoring the properties of Ti-Al-C carbides by controlling the microstructures. Study of the microstructure of TiAl-containing Ti$_2$AlC showed that Ti$_2$AlC formed preferentially at TiAl twins. Similar results showed[72] that the main phase among combustion products changed from Ti$_3$AlC$_2$ to Ti$_2$AlC if TiC was added in small quantities. The amount of Ti$_2$AlC increased with decreasing combustion temperature. The upper limit to the combustion temperature for the formation of Ti$_3$AlC$_2$ was much higher than that for Ti$_2$AlC. The latter was difficult to form when the combustion temperature was above 1600C, while Ti$_3$AlC$_2$ could still be formed.

*Table 30. Moduli of $M_2InC$ phases*

| Phase | B (GPa) | G (GPa) | E (GPa) | Poisson Ratio |
|-------|---------|---------|---------|---------------|
| $Zr_2InC$ | 137 | 99 | 239 | 0.208 |
| $Hf_2InC$ | 168 | 109 | 270 | 0.232 |
| $Ta_2InC$ | 248 | 157 | 390 | 0.23 |

A dense 70μm-thick coating of this material was deposited onto Inconel 625 by cold-spraying[73]. The ball-on-disk wear behaviour at 25 and 600C was studied, showing that the coefficient-of-friction and volume-loss at 600C were reduced by 21 and 40%, respectively, due to lubrication by an oxide layer which formed at the higher temperature. For a load of 7000μN at room temperature, the coating exhibited an elastic modulus of 273GPa; as compared to the elastic modulus of 191GPa which was observed at 300C. Room-temperature nano-scratching under 7000μN revealed a brittle behaviour, with fracture, chipping and debris-formation. High-temperature (300C) scratching revealed ductile behaviour, with plowing, cutting and zero debris-formation. The wear-volume loss was some orders of magnitude higher for 8N scratch loading.

Materials having compositions which were close to those of MAX phases were created[74] in Ti–Al–C powder mixtures during self-propagating high-temperature synthesis. In the first stage, the main reaction was TiC formation and this exothermic reaction then led to the formation of TiC crystals which were surrounded by Ti-Al melt. The carbide then dissolved in the surrounding melt, with subsequent crystallization of $Ti_2AlC$. The final product was based upon $Ti_2AlC$ which contained less than 20wt% of TiAl and 2wt% of TiC. The structure consisted of lamellar $Ti_2AlC$ grains which were surrounded by a TiAl matrix. The microhardness was between 4.0 and 4.5GPa, and reflected the microhardness of the $Ti_2AlC$. Specimens were prepared[75] by the cold-pressing of titanium, aluminium and silicon carbide and subsequent high-temperature sintering under vacuum. The $Ti_2AlC$ was a result of the interaction of $Ti_3Al$ with carbon atoms that were released during the reaction of SiC with titanium. Complete sintering occurred at above 1300C. In other cases, there were also individual grains of TiC, SiC and $Ti_3Al$. Samples which were sintered at 1400C were 2-phase composites in which $Ti_2AlC$ and $Ti_5Si_3C_x$ were evenly distributed. There were no other phases in this case. The Vickers microhardness of the 2-phase grains was 7.2GPa and the hardness ranged from 3.0 to 9.0GPa due to inhomogeneity of the microstructure.

*Table 31. Hardness of $M_2InC$ phases*

| Phase | Hv (GPa) |
|-------|----------|
| $Zr_2InC$ | 1.05 |
| $Hf_2InC$ | 3.45 |
| $Ta_2InC$ | 4.12 |

Textured lamellar composites exhibited[76] a compressive strength of about 2GPa, a fracture toughness of 8.5MPam$^{0.5}$ parallel to the c-axis, a flexural strength of 735MPa parallel to the c-axis and a hardness of 7.9GPa parallel to the c-axis. When bulk samples were prepared[77] by aqueous gel-casting and aluminium-rich pressureless sintering, a synergistic effect of a porous reaction layer and open channels resulted in a sharp decrease in the density and mechanical properties of carbon-rich pressureless sintered samples. On the other hand, aluminium-rich pressureless sintering provided strong protection for the sintering of gel-cast green bodies and led to a density, hardness, flexural strength and fracture toughness of 3.98g/cm$^3$, 4.29GPa, 341MPa and 4.83MPam$^{0.5}$, respectively.

The effect of a boronizing treatment upon the friction and wear behaviours of $Ti_2AlC$ in contact with alumina was determined[78] by means of powder-pack cementation (1350C, 8h). During boronizing in the presence of oxygen, the boron atoms reacted with the $Ti_2AlC$ to form $TiB_2$, $TiC$ and $Al_2O_3$. An enhanced friction coefficient was attributed to a very high hardness and to the load-carrying ability $TiB_2$ and the boronizing increased the surface hardness from 3.0 to 24.25GPa. High-velocity oxy-fuel was used[79] to spray $Ti_2AlC$, yielding coatings which consisted mainly of $Ti_2AlC$ plus inclusions of $Ti_3AlC_2$, $TiC$ and Al-Ti alloy. The fraction of $Ti_2AlC$ in coatings which were sprayed using powder with a size of 38μm increased with decreasing spray-power. Coarser (56μm) powder yielded a higher volume fraction of $Ti_2AlC$ in the coatings; at the cost of increasing porosity. There was a preferred crystal orientation of the coatings, with the $Ti_2AlC$ (00•l) planes being aligned with the substrate surface. Indentation tests indicated a hardness of 3 to 5GPa when coatings were prepared using a powder size of 38μm.

Thin films were prepared[80] via the heat treatment of Ti-Al-C multilayers which had been made by magnetron sputtering. The $Ti_2AlC$ phase formed below 850C, while $Ti_3AlC_2$ started to form at above 850C and attained its greatest phase purity at 950C. The indentation hardness depended upon the annealing temperature (figure 23). The measurements indicated that $Ti_2AlC$, with a value of 10.2GPa, was harder than the

$Ti_3AlC_2$, with a value of 3.2GPa. The grain size of the $Ti_2AlC$ was smaller and so the hardness was much higher than that of the $Ti_3AlC_2$, due to the grain boundaries acting as barriers to dislocation motion. The hardness results also reflected the indentation-size effect, in which the hardness is higher for small penetration depths

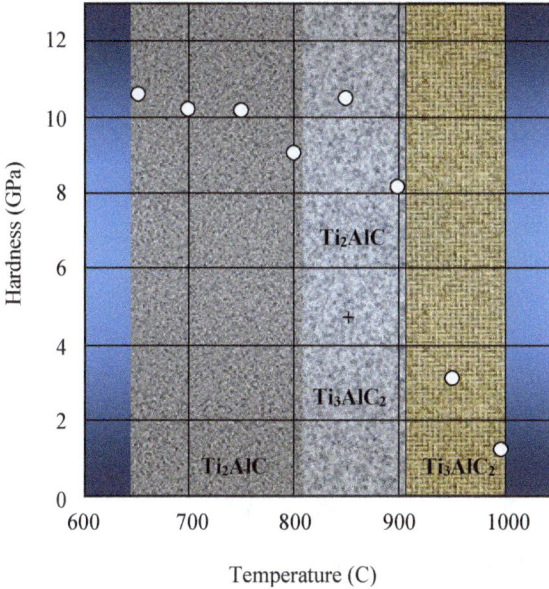

*Figure 23. Hardness of $Ti_2AlC$ thin films as a function of the annealing temperature*

The effect of boronizing upon the behaviour of $Ti_2AlC$ in contact with alumina was studied[81] by carrying out powder-pack cementation (1350C, 8h). During the boronizing, in the presence of some oxygen, boron atoms diffused and reacted with the $Ti_2AlC$ to produce $TiB_2$, $TiC$ and $Al_2O_3$. An improved friction coefficient was attributed to a very high hardness and to the load-carrying ability of the $TiB_2$. The boronizing increased the surface hardness of $Ti_2AlC$ from 3.0 to 24.25GPa. Plasma-nitriding was also used[82] to improve the surface hardness of $Ti_2AlC$ because of its low natural hardness of about 3GPa. A continuous coating was formed by plasma-nitriding (800C, 4h) the $Ti_2AlC$ in a mixed atmosphere of 75vol%$N_2$ and 25vol%$H_2$. The resultant surface layer was about

6μm thick and comprised TiN plus a small amount of AlN. The strong coherent coating improved the surface hardness.

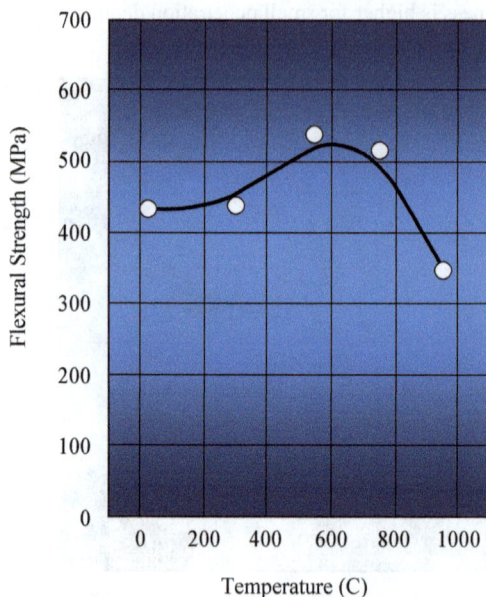

*Figure 24. Flexural strength of Ti₂AlC as a function of temperature*

Bulk $Ti_2AlC$ was prepared[83] via self-propagating high-temperature combustion and pseudo hot-isostatic pressing. With increasing temperature, lattice defects contributed to a decreasing phonon thermal conductivity and the electrical resistivity increased linearly from room temperature to 900C. The room-temperature flexural strength, compressive strength, fracture toughness, work-of-fracture and Vickers hardness were 606MPa, 1057MPa, $6.9MPam^{0.5}$, $158J/m^2$ and 4.7GPa, respectively. With increasing temperature, the flexural (figure 24) and compressive strengths (figure 25) remained almost unchanged in the zone of brittle failure, but decreased markedly when plastic deformation occurred. The brittle-ductile transition temperature under flexure (900 to 950C) was higher than that under compression (700 to 800C). In similarly-treated samples[84] which contained non-stoichiometric $Ti_2AlC_{0.69}$ with a grain size of about 6μm, the Vickers hardness was 5.8GPa, the flexural strength was 432MPa, the compressive strength was 1037MPa and the fracture toughness was 6.5MPam$^{0.5}$. No indentation cracks were observed. Solid-liquid reaction synthesis and simultaneous densification was used[85] to prepare samples from a stoichiometric mixture of titanium, aluminium and graphite powders. The room-temperature electrical conductivity was 4.42 x 10$^6$S/m, and increased with decreasing temperature, while the Vickers hardness was 2.8GPa. As above, no indentation cracks were observed and the material was deemed to be damage-tolerant. The flexural strength was 275MPa and, as above, the fracture toughness was 6.5MPam$^{0.5}$.

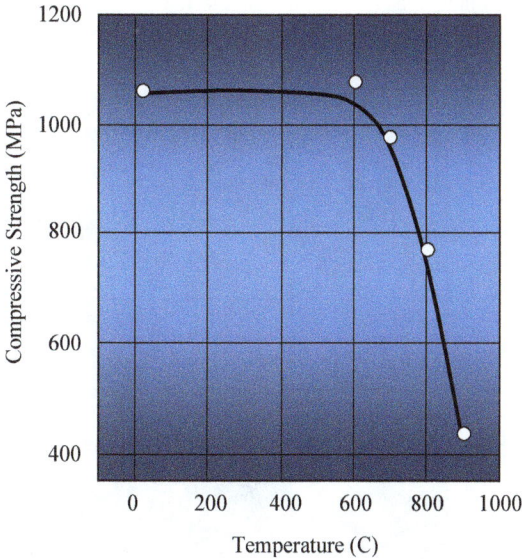

Figure 25. Compressive strength of Ti$_2$AlC as a function of temperature

Table 32. Comparison of the properties of Ti$_2$AlC and its analogues

| Property | Ti$_3$SiC$_2$ | Ti$_2$AlC | Ti$_3$GeC$_2$ | Ti$_3$Si$_{0.5}$Al$_{0.5}$C$_2$ |
|---|---|---|---|---|
| Density (g/cm$^3$) | 4.50 | 4.0 | 5.29 | 4.35 |
| Young's modulus (GPa) | 343 | 277.6 | 347 | 322 |
| Shear modulus (GPa) | 143.8 | 118.8 | 145 | 136.8 |
| Bulk modulus (GPa) | 185.6 | 139.6 | 174 | 166 |
| Poisson ratio | 0.192 | 0.169 | 0.196 | 0.176 |

Machinable $Ti_2AlC$ was prepared[86] by spark plasma sintering of $Ti_2AlC$ powders. The relative density was 98.6% and the Vickers hardness was 4.3GPa for sintered (1250C, 300s, 20MPa, vacuum) samples. The Vickers hardness of sintered samples increased with increasing sintering temperature up to 1250C, before decreasing and decomposing at above 1350C. The lattice parameters of $Ti_2AlN$ and $Ti_2AlC$ powders were measured[87] as a function of pressure at up to 50GPa, with no phase transformation being observed. The compressibilities along the c-axis were larger than those along the a-axis, and the bulk modulus of $Ti_2AlC$, 186GPa, was some 10% higher than that, 169GPa, of $Ti_2AlN$. Investigations were made[88], at 5 to 300K, of the elastic properties of $Ti_2AlC$, $V_2AlC$, $Cr_2AlC$ and $Nb_2AlC$. The Young's and shear moduli were of the order of 270 and 120GPa, respectively, and the phases were therefore relatively stiff.

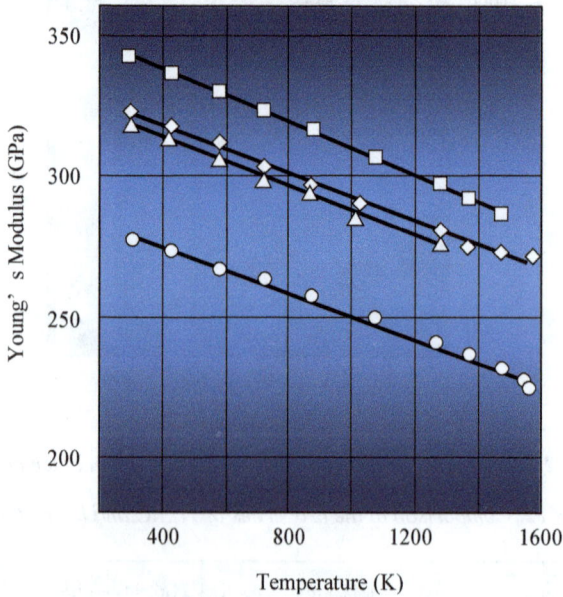

Figure 26. Young's modulus of $Ti_2AlC$ and its analogues as a function of temperature. Squares: $Ti_3SiC_2$, diamonds: $Ti_2AlC$, triangles: $Ti_3GeC_2$, circles: $Ti_3SiAlC_2$.

The temperature dependences of the Young's and shear moduli (table 32, figures 26 and 27) of polycrystalline $Ti_3SiC_2$, $Ti_2AlC$, $Ti_3GeC_2$ and $Ti_3Si_{0.5}Al_{0.5}C_2$ were determined[89] by means of resonant ultrasound spectroscopy at 300 to 1573K. In the case of isostructural

312 phases, the longitudinal and shear sound velocities decreased in the order: $Ti_3GeC_2$ > $Ti_3Si_{0.5}Al_{0.5}C_2$ > $Ti_3SiC_2$ > $Ti_3GeC_2$. These solids were relatively stiff, with the room-temperature Young's moduli ranging from between 340 and 277GPa for $Ti_2AlC$ to 340GPa for $Ti_3SiC_2$ while the corresponding shear moduli ranged from 119 to 144GPa. The Poisson ratio was about 0.19. The Young's and shear moduli decreased linearly and slowly with increasing temperature in all of these materials. Damping was attributed to the interaction of dislocation segments with ultrasound waves. Dense polycrystalline samples were prepared[90] via the spark plasma sintering (1100C, 1h, 30MPa) of titanium, aluminium and carbon powders in the molar ratio of 2:1.2:1. The relative density was 99.8%, with lattice parameters of a = 0.3058nm and c = 1.3649nm and a Vickers hardness (1N) of 4GPa.

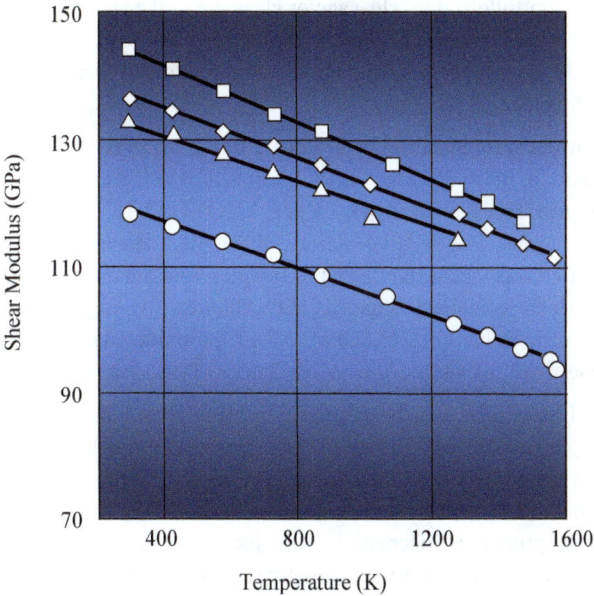

*Figure 27. Shear modulus of $Ti_2AlC$ and its analogues as a function of temperature. Squares: $Ti_3SiC_2$, diamonds: $Ti_3SiAlC_2$, triangles: $Ti_3GeC_2$, circles: $Ti_2AlC$.*

## Toughness

It has been demonstrated[91] that this metallo-ceramic can possess a bone-like hierarchical microstructure and can fail along zig-zag fracture surfaces. It can also exhibit repeated full strength and toughness recovery at room temperature, with the high-temperature healing reaction involving the local formation of dense brittle alumina within the crack. Full recovery of the fracture toughness depends upon the healed zone thickness and upon the process-zone size formed in the alumina reaction product.

A study was made[92] of the mechanical deformation of the single-crystal phase by means of the compression testing of micropillars having a range of crystallographic orientations. This showed that, depending upon the crystallographic orientation, the micropillars underwent either non-classical (non-Schmid) crystallographic slip, non-classical crystallographic slip followed by cleavage or cleavage in the absence of any appreciable crystallographic slip. The non-classical crystallographic slip was a result of the strong dependence of the crystallographic slip upon both the resolved shear stress and the stress normal to the slip plane.

Pure $Ti_2AlC$ powder was prepared[93] by heating (1300C, 4h) an Al:Ti:TiC powder mixture with the molar ratio, 1:1:0.75. The $Ti_2AlC$ compacts which were used for pressureless sintering contained 5mass% of $Al_2O_3$, $Y_2O_3$, MgO, CaO or $TiO_2$, and were heated (1400C, 2h) in argon. The relative density of $Ti_2AlC$ samples which were sintered without an additive was 94.2%, while that of samples with 5mass% of $Al_2O_3$ was 96.0%. The samples which contained 5mass%CaO exhibited the highest fracture strength (460MPa) and Vickers hardness (5.8GPa). All of the oxide additives led to a reduced fracture toughness, while that of additive-free sintered $Ti_2AlC$ samples was $5.2MPam^{0.5}$.

The tribological behaviour was investigated[94] by nano-scratching and nano-indentation. Samples were prepared by hot isostatic pressing, giving a roughness of $0.002\mu m$. The nano-scratching tests were performed using low loads: 2000 to $10000\mu N$. The results indicated that the mechanical properties decreased with increasing load and dwell-time. The tribological properties decreased with increasing coefficient of friction and increasing load. The deformation microstructure that developed during nano-indentation was characterized[95] by scanning probe microscopy and transmission electron microscopy. In order to investigate the plastic anisotropy, nano-indentation measurements were performed on grains with the normal parallel to $<33\bullet2>$, $<00\bullet1>$ and $<11\bullet0>$. Basal slip, $\{00\bullet1\}<11\bullet0>$, was the predominant deformation mechanism for all of the indentation directions. When nano-indenting along $<00\bullet1>$ and $<11\bullet0>$, non-basal slip occurred beneath the indenter and the slip system of non-basal dislocations was $(1\bar{2}\bullet6)[1\bar{2}\bullet1]$. Fine-scale kink-bands formed beneath the residual impression and the formation of the kink-band was associated with delamination in the form of micro-cracking along the basal plane. This suggested that delamination played an important role in kink-band formation. In similar work, polycrystalline samples were prepared[96] by spark plasma sintering of high-purity elements and annealing (1500C, 24h). The nano-indentation behavior was studied here by using grains with their surface normals parallel to <33•2>, <44•1>, <00•1>and <11•0>. During loading, pop-in events occurred for all of the indentation directions, but the critical load which was associated with the first pop-in, for indentation directions parallel to <11•0> and <00•1>, was higher than that for <33•2> and <44•1>. The surface topography surrounding the residual indents indicated that basal slip again dominated plastic deformation during indentation, and that the nucleation of basal dislocations was a major barrier to the onset of plastic deformation beneath an indenter.

This material also exhibits an excellent resistance to irradiation, and a study[97] was aimed at investigating the effect of high temperatures upon the evolution of irradiation damage. Samples were bombarded with 1.1MeV $C^{4+}$ ions, at 298 or 873K, using fluences ranging from 2 x $10^{15}$ to 6 x $10^{16}$/cm$^2$. Significant cracking occurred along grain boundaries in the case of bombardment at 298K, but not after bombardment at 837K. No amorphization occurred, and some face-centered-cubic phases formed during high-dose bombardment at 298K. Nano-indentation testing revealed the occurrence of marked hardening after bombardment at high temperatures, thus suggesting the presence of irradiation-induced defects. A study[98] was also made of the defects which formed in polycrystalline $Ti_3SiC_2$ and $Ti_2AlC$ during neutron irradiation at up to 0.1dpa at 350 or 695C, and at up to 0.4dpa at 350C. Black spots were observed in both materials following irradiation to 0.1 or 0.4dpa at 350C. Following irradiation to 0.1dpa at 695C, small basal dislocation loops having a Burgers vector of ½[00•1] were observed in both materials, with loop diameters of 9 and 10nm in $Ti_3SiC_2$ and $Ti_2AlC$ samples, respectively. At 1 x $10^{23}$ loops/m$^3$, the dislocation loop density in $Ti_2AlC$ was some 1.5 orders of magnitude greater than that of the 3 x $10^{21}$ loops/m$^3$ in $Ti_3SiC_2$. Following irradiation at 350C, extensive microcracking was observed in $Ti_2AlC$ but not in $Ti_3SiC_2$. It was noted that these MAX phases were clearly more neutron radiation tolerant than were TiC and $Al_2O_3$, and that $Ti_3SiC_2$ was a more promising MAX-phase for high-temperature nuclear applications than was $Ti_2AlC$. In a further study, 1MeV Au ions were used[99] to bombard $Ti_2AlC$ to fluences ranging from 1 x $10^{14}$ to 2 x $10^{16}$/cm$^2$ at room temperature. The nanolayered structures were partially disturbed by the lowest fluence, 1 x $10^{14}$/cm$^2$, with a phase from the original hexagonal close-packed structure to a γ-$Ti_2AlC$ structure. With increasing fluence, the presence of more extended defects led to symmetry variation and to the transformation of a possible δ-phase structure. At the highest fluence, 2 x $10^{16}$/cm$^2$, face-centered-cubic structures formed from the original hexagonal close-packed structure, without amorphization. This indicated a marked tolerance of irradiation-induced amorphization.

The high (up to 4700/s) strain-rate compressive response was studied using the split Hopkinson pressure bar technique[100]. An optimized specimen geometry ensured the occurrence of dynamic equilibrium, and minimized dispersion of the transmitted pulse. *In situ* high-speed imaging facilitated the identification of real strains, free from macroscopic crack artefacts. The results revealed the occurrence of appreciable inelastic deformation and strain softening before fracture; even at very high strain-rates. Post-fracture microstructures revealed the simultaneous coexistence of kink bands and delamination, grain pull-outs and transgranular cracks due to the high strain-rate loading. These characteristics, which also applied to quasi-static loading, were suggested to be the cause of the exceptional damage-tolerance of this material and of the high-rate kink-banding which was observed in MAX phases.

A study was made[101] of the response of fully dense, and 10vol% porous, polycrystalline samples to uniaxial compression and indentation by a hemispherical nano-indenter. The kinking-induced behavior could be described by four interrelated loading parameters: the stress, the non-linear strain, the stored non-linear energy per unit volume and the dissipated energy per unit volume. These parameters could all be deduced from stress-strain curves. The experimental data described were in excellent agreement with a proposed model in which the dissipated energy per unit volume scaled with the square of the stress, while the stored non-linear energy per unit volume scaled with the 3/2 power of the non-linear strain. Both relationships applied over a very large stress-range. Further analysis of the results indicated that the dislocation density at a maximum stress of 350MPa was about $10^{13}/m^2$. The critical resolved shear stress on the basal planes was also estimated to be about 22MPa.

## Ti₃AlC

### *Elastic constants*

A first-principles study was made[102] of the elastic properties. The absence of a band-gap at the Fermi level and the finite value of the density-of-states at the Fermi energy indicated a metallic behaviour for this material. The elastic constants were $C_{11}$ = 356GPa and $C_{44}$ = 157GPa, with the bulk, shear and Young's moduli being 156, 151 and 342GPa, respectively.

## Ti$_3$AlC$_2$

### Hardness

Samples were prepared[103] from powder mixtures of titanium and carbon by using spark plasma sintering. The highest-purity (97.23wt%) material was produced when the ratio of the Ti:Al:C powders was 3:1.2:2 and were heated at 1300C for 1h. The product had a compact and uniform lath-like structure with a length/thickness ratio of 3 to 5. The excellent mechanical properties included a Vickers hardness, bending strength and fracture toughness of 5.26GPa, 500MPa and 7.35MPa·m$^{0.5}$, respectively. The fracture morphologies indicated that extrusion and kinking occurred in the layered structure. These two forms of energy dissipation led to the bending strength and fracture toughness being higher than those of traditional ceramics. A study[104] of the slip-casting and pressureless sintering which was used to prepare large and complex Ti$_3$AlC$_2$ components showed that the sintering temperature and embedding powder controlled the properties of the sintered material. Samples which were sintered (1450C, 1.5h) with Al$_4$C$_3$ embedding powder exhibited the best properties, such as a 95.3% relative density, a hardness of 4.18GPa, a thermal conductivity of 29.11W/mK and an electrical resistivity of 0.39μΩm. The mechanical properties of high-density (4.27g/cm$^3$, 1% porosity) material, based upon hot-pressed (30MPa) nanolaminated Ti$_3$AlC$_2$ (89%Ti$_3$AlC$_2$, 6%TiC, 5%Al$_2$O$_3$), were investigated[105]. At room temperature, the samples exhibited a Vickers (5N) microhardness of 4.6GPa, a Vickers (50N) hardness of 630MPa, a Young's modulus of 140GPa, a fracture toughness of 10.2MPam$^{0.5}$, a compressive strength of 700MPa and a bending strength of 500MPa. Due to surface oxidation, defect self-healing occurred and the bending strength of 22% porous Ti$_3$AlC$_2$, following exposure to air (3h, 600C) increased as compared with that at 20C. The porous material also resisted high-temperature creep while, following exposure (600C, 3h) to H$_2$, its bending strength was reduced by 5%. Thermogravimetry and differential thermal analysis were used[106] to study the resistance to oxidation in air of high-density samples of Ti$_3$AlC$_2$, Ti$_2$AlC and Ti$_2$Al(C$_{1-x}$N$_x$) solid solutions. The Ti$_3$AlC$_2$ was more stable than the Ti$_2$AlC and Ti$_2$Al(C$_{1-x}$N$_x$) and, as x increased to 0.75, the oxidation resistance decreased. Material which contained 89wt%Ti$_3$AlC$_2$, and had a relative density of 99%, exhibited a bending strength of 500MPa, a compressive strength of 700MPa, a fracture toughness of 10.2MPam$^{0.5}$, a Vickers hardness of 4.6GPa and a Young's modulus of 149.4GPa.

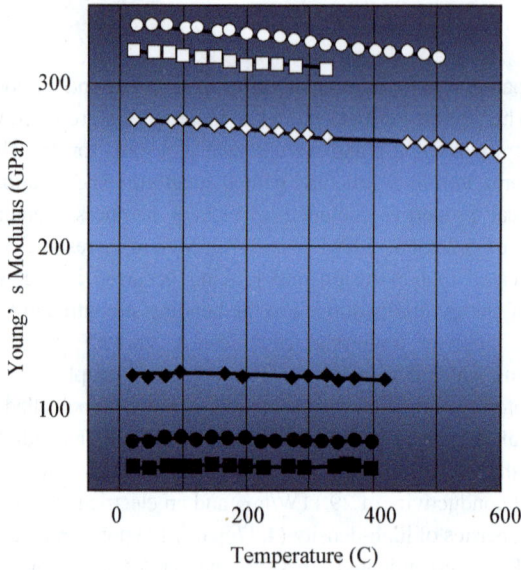

*Figure 28. Young's moduli of Ti$_2$AlC and Ti$_3$AlC$_2$ as a function of temperature. White circles: 62.8wt%Ti$_2$AlC, 30.1wt%Ti3AlC$_2$, 2.8wt%TiC, 4.3%Al$_2$O$_3$, white squares: 98.7wt%Ti$_2$AlC, 1.3wt%TiC, white diamonds: 88.3wt%Ti$_2$AlC, 10.2wt%Ti$_3$AlC$_2$, 1.5wt%TiC, black diamonds: 90.3wt%Ti$_3$AlC$_2$, 9.7wt%TiC, black circles: 86.8wt%Ti$_3$AlC$_2$, 13.2wt%TiC, black squares: 38.6wt%Ti$_3$AlC$_2$, 41.4wt%Ti$_2$AlC, 20.0wt%TiC.*

Polycrystalline bulk Ti$_3$AlC$_2$ was hot-pressed (800-1600C, 25MPa) from TiC$_{0.6}$ and aluminium powder mixtures[107]. A small amount of Ti$_2$AlC secondary phase was present, together with Ti$_3$AlC$_2$, below 1400C. With increasing hot-pressing time at 1250C, the amount of non-reacted TiC$_x$ gradually decreased while Ti$_3$AlC$_2$ appeared to predominate. Almost fully dense Ti$_3$AlC$_2$ was produced as the hot-pressing time was increased at 1250C. Fully dense pure Ti$_3$AlC$_2$ could be produced by hot-pressing at above 1400C. The Vickers hardness (10N) of bulk Ti$_3$AlC$_2$ was 3.5 to 6GPa and the maximum flexural strength was over 900MPa. Samples of Ti$_3$AlC$_2$-based material were prepared[108] by first creating a reactive powder, which consisted mainly of Ti$_2$AlC, by pressureless sintering (1500C, 600s) in argon at a heating-rate of 20C/min. The powder was then hot-pressed

(1300C, 1h) to yield bulk material. This second step densified the powder and promoted phase transformation from $Ti_2AlC$ to $Ti_3AlC_2$. The resultant specimens comprised $Ti_3AlC_2$, $Ti_2AlC$ and some TiC. The density was $4.26g/cm^3$, with a flexural strength of 664.4MPa, a Vickers hardness of 6.4GPa and a fracture toughness of $9.9MPam^{0.5}$.

*Table 33. Properties of $Ti_3Si(Al)C_2$, and $Ti_3AlC_2$*

| Phase | Grain Size | E (GPa) | Vickers Hardness (GPa) | Flexural Strength (MPa) |
|---|---|---|---|---|
| $Ti_3Si(Al)C_2$ | fine | 336 | 4.02 | 458.5 |
| $Ti_3Si(Al)C_2$ | coarse | 335 | 3.28 | 306.4 |
| $Ti_3AlC_2$ | fine | 304 | 3.46 | 320.3 |
| $Ti_3AlC_2$ | coarse | 301 | 2.22 | 169.2 |

### *Elastic constants*

Polycrystalline samples of $Ti_2AlC$ and $Ti_3AlC_2$ were prepared[109] from metal powder by hot-pressing at temperatures of 1100 or 1300C, and the Young's modulus was measured between 25 and 600C (figure 28). In the case of $Ti_2AlC$, there was no apparent change in the Young's modulus as a function of temperature; with the values lying between 50 and 125GPa. This was much lower than previously reported and was attributed to a high (23 to 34%) degree of porosity. In these samples, the composition did not affect the elastic properties. In the case of $Ti_3AlC_2$, the Young's modulus decreased by just 10% between 25 and 600C. The Young's modulus of about 325GPa was similar to previously reported values. A sample having the lower Young's modulus of about 275GPa had a multi-phase composition in which 41% of the material was occupied by $Ti_2AlC$ having a lower (260GPa) modulus and density (95%). Compression tests were performed[110] on single-crystal micropillars having a range of crystallographic orientations. The stress-strain response in compression exhibited 3 characteristic stages: the stress-strain response was initially linear but, with continued deformation, the response deviated from linearity; indicating the onset of yielding. Following yielding, the response exhibited a short hardening period that eventually saturated. The critical resolved shear stress differed for different crystallographic orientations, as in the case of $Ti_2AlC$. The crystallographic slip of $Ti_3AlC_2$ thus did not obey the Schmid law that the critical resolved shear stress should be independent of orientation. The stress-strain response of two pairs of micropillars of

$Ti_3AlC_2$ and $Ti_2AlC$, having orientations that resulted in essentially the same maximum Schmid factor for basal slip, indicated a maximum factor of about 0.49 for a $Ti_3AlC_2$ micropillar and about 0.5 for a $Ti_2AlC$ micropillar. Both $Ti_3AlC_2$ and $Ti_2AlC$ exhibited a non-classical mechanism of crystallographic slip. Due to their differing atomic stackings, the frictional coefficient for basal slip in $Ti_3AlC_2$ was some 28.6% smaller than that in $Ti_2AlC$. The intrinsic critical resolved shear stress for basal slip in $Ti_3AlC_2$ was about 42% higher than that in $Ti_2AlC$. In general, $Ti_3AlC_2$ was stiffer than $Ti_2AlC$ and, for a given grain size, $Ti_3AlC_2$ had a slightly higher compressive strength than $Ti_2AlC$. A hysteresis of MAX phases during loading-unloading cycles is associated with the occurrence of reversible plastic flow due to the continuous build-up and relaxation of large incompatibility stresses.

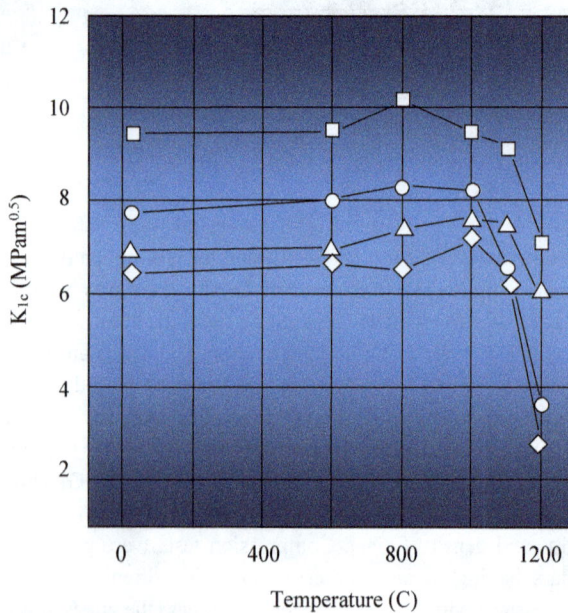

Figure 29. Fracture toughness of fine-grained $Ti_3Si(Al)C_2$. Squares: coarse-grained $Ti_3Si(Al)C_2$, circles: fine-grained $Ti_3AlC_2$, triangles & diamonds: coarse-grained $Ti_3AlC_2$.

*Table 34. Calculated elastic coefficients (GPa) for Ti$_2$Al(C$_x$N$_{1-x}$)*

| x | C$_{11}$ | C$_{33}$ | C$_{44}$ | C$_{66}$ | C$_{12}$ | C$_{13}$ | Poisson Ratio |
|---|---|---|---|---|---|---|---|
| 0.000 | 314.94 | 292.58 | 128.57 | 122.44 | 71.87 | 94.96 | 0.743 |
| 0.062 | 315.04 | 290.06 | 128.15 | 123.02 | 69.16 | 93.07 | 0.752 |
| 0.125 | 314.69 | 288.32 | 127.84 | 123.64 | 67.50 | 91.65 | 0.759 |
| 0.188 | 313.41 | 286.98 | 127.26 | 123.51 | 66.26 | 90.00 | 0.765 |
| 0.250 | 310.73 | 285.45 | 126.92 | 122.99 | 64.36 | 89.33 | 0.769 |
| 0.312 | 307.31 | 282.14 | 125.76 | 122.33 | 62.40 | 88.44 | 0.772 |
| 0.375 | 303.49 | 279.61 | 124.70 | 121.42 | 60.57 | 88.20 | 0.773 |
| 0.438 | 300.31 | 277.63 | 123.77 | 120.81 | 59.20 | 87.68 | 0.774 |
| 0.500 | 297.78 | 275.63 | 122.34 | 119.71 | 58.64 | 86.18 | 0.776 |
| 0.562 | 297.83 | 275.09 | 120.89 | 119.11 | 59.90 | 83.24 | 0.780 |
| 0.625 | 298.71 | 274.85 | 119.33 | 118.56 | 61.92 | 79.46 | 0.785 |
| 0.688 | 300.42 | 275.20 | 117.94 | 117.86 | 64.82 | 75.45 | 0.789 |
| 0.750 | 301.98 | 274.99 | 116.22 | 117.56 | 67.03 | 72.18 | 0.791 |
| 0.812 | 303.48 | 274.63 | 114.54 | 117.64 | 68.64 | 69.21 | 0.794 |
| 0.875 | 304.92 | 274.14 | 112.78 | 118.38 | 68.79 | 66.81 | 0.798 |
| 0.938 | 305.64 | 272.54 | 110.68 | 119.00 | 67.84 | 65.02 | 0.801 |
| 1.000 | 305.34 | 269.46 | 108.62 | 119.16 | 66.69 | 63.26 | 0.804 |

## Toughness

The effects of grain-size, notch-width and testing temperature upon the fracture toughness of Ti$_3$Si(Al)C$_2$ and Ti$_3$AlC$_2$, were investigated[111], revealing fracture toughness values ranging from 6.4 to 7.2MPam$^{0.5}$ for Ti$_3$Si(Al)C$_2$ and from 7.6 to 10.2MPam$^{0.5}$ for Ti$_3$AlC$_2$. The critical notch-width for valid fracture toughness measurement was about 250μm. For a given notch-width and testing temperature, the fracture toughness of coarse-grained samples was higher than that of fine-grained samples (table 33). The high-temperature fracture toughness values of Ti$_3$Si(Al)C$_2$ and Ti$_3$AlC$_2$ (figure 29) were such that they exhibited a similar trend in that the measured toughness was nearly constant when the testing temperature was below the ductile-brittle transition temperature. It then decreased and finally fell sharply above the ductile-brittle transition temperature. The decrease in the fracture toughness was attributed to the decrease in elastic modulus at temperatures above the ductile-brittle transition temperature.

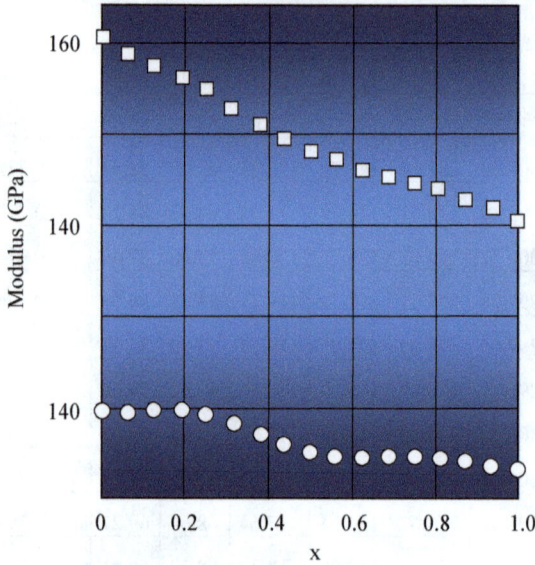

*Figure 30. Bulk (squares) and shear (circles)*
*moduli of Ti₂Al(CₓN₁₋ₓ) as a function of x*

### Creep

Study of the deformation mechanisms which were involved in the tensile creep of specimens that were deformed at 900C to a final strain of 7.5% showed[112] that the tensile-creep strain-rate obeyed a Norton-type law with a power exponent of about 2. This implied that the creep was controlled by grain-boundary sliding. The microstructure was very heterogeneous and exhibited both grains with no dislocations and grains which were highly defected. The intragranular deformation involved 3 distinct microstructural features. That is, dislocations which were largely confined to basal planes and organized into hexagonal networks, together with numerous stacking faults and lenticular non-planar defects.

## Ti$_2$Al(C,N)

### *Elastic constants*

Elastic constants (table 34) and moduli (figures 30 and 31) were calculated[113], using first-principles density functional theory and 4 x 4 x 1 super-cell models, for Ti$_2$Al(C$_x$N$_{1-x}$) where x ranged from 0 to 1. The properties did not vary linearly as a function of x. The *c* lattice constant was almost constant when x was less than 0.5, while the *a* lattice constant increased linearly. When x was greater than 0.5, the *c* lattice constant started to increase, while the rate-of-increase of the *a* lattice constant slowed. When x was between 0.5 and 0.85, the elastic constants and mechanical properties reflected a complicated interplay between the structures and properties of the solid solutions. Non-linear variations in the properties were attributed to details of the electronic structures and bonding between nitrides and carbides.

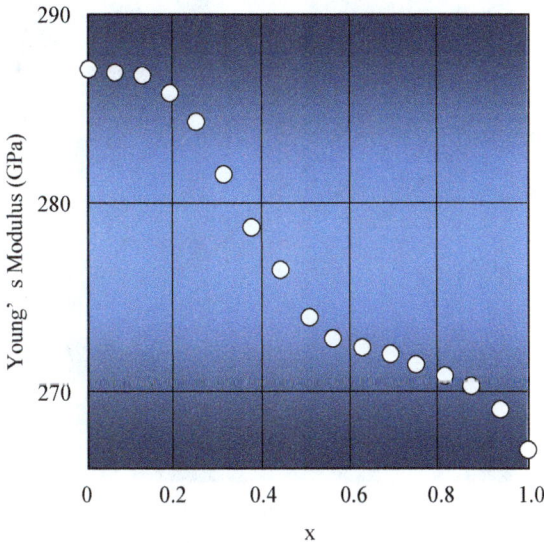

*Figure 31. Young's modulus of Ti$_2$Al(C$_x$N$_{1-x}$) as a function of x*

Materials Research Forum LLC
https://doi.org/10.21741/9781644901274

## $Ti_2AlN$

### *Creep*

High-density high-purity bulk material was prepared[114] at 1400C by means of the field-assisted sintering of titanium, aluminium and TiN powders. The aluminum content appeared to play a pivotal role in obtaining high phase-purity, with the optimum starting molar ratio being 1:1.02:1 for the titanium, aluminium and TiNi. The elastic modulus and hardness were determined by micro-indentation testing at room temperature, and creep tests were performed in air at 900 to 1200C under constant applied stresses ranging from 20 to 100MPa. The elastic moduli and hardness values of samples with an initial aluminium ratio of 1.02, sintered for 0.25 or 0.5h, were 252 and 7 to 11GPa or 254 and 7 to 10GPa, respectively. Sand-blasting with 2bar compressed air for 600s led to mass losses of 0.73 or 0.66wt%, respectively, for the same samples. The creep results (figures 32 and 33) were comparable to those for other carbide phases and the creep behaviour, with an activation energy of 445 to 537kJ/mol and a stress exponent of 1.5 to 1.9, was attributed to control by dislocation motion and grain-boundary sliding.

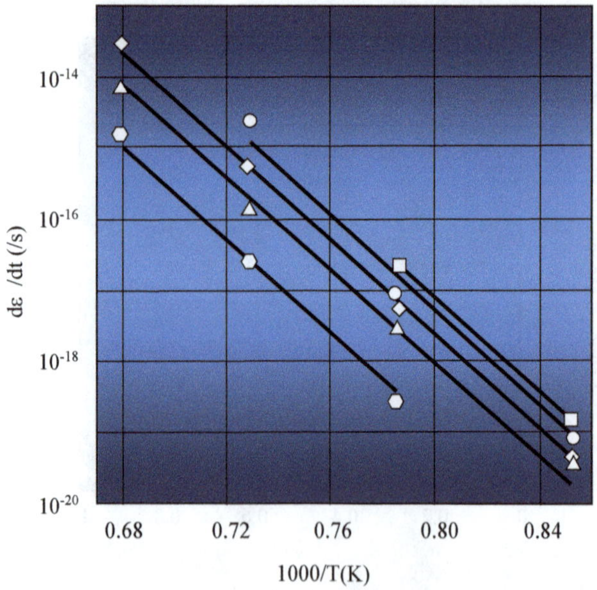


*Figure 32. Creep rate of $Ti_2AlN$, sintered for 0.25h, as a function of temperature. Hexagons: 20MPa, triangles: 40MPa, diamonds: 60MPa, circles: 80MPa, squares: 100MPa.*

*Table 35. Calculated elastic constants and shear anisotropy factors of Ti₂AlN andCr₂AlC*

| Phase | $C_{11}$ (GPa) | $C_{12}$ (GPa) | $C_{13}$ (GPa) | $C_{33}$ (GPa) | $C_{44}$ (GPa) | A |
|-------|------|------|------|------|------|------|
| Ti₂AlN | 303.57 | 67.78 | 91.37 | 289.95 | 135.43 | 1.3187 |
| Cr₂AlC | 344.48 | 67.5 | 95.4 | 332.58 | 153.24 | 1.1125 |

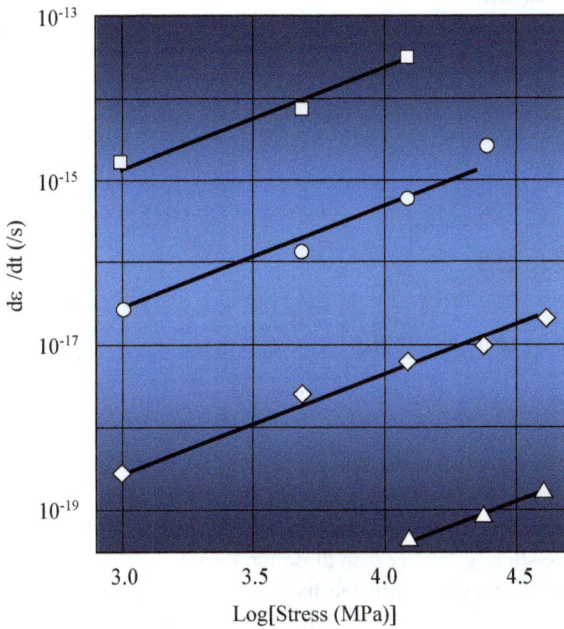

*Figure 33. Creep rate of Ti₂AlN, sintered for 0.25h, as a function of stress. Squares: 1200C, circles: 1100C, diamonds: 1000C, triangles: 900C.*

*Table 36. Calculated moduli and Poisson ratios of Ti₂AlN andCr₂AlC*

| Phase | B (GPa) | G (GPa) | E (GPa) | Poisson Ratio |
|-------|------|------|------|------|
| Ti₂AlN | 155.33 | 119.98 | 286.25 | 0.192 |
| Cr₂AlC | 170.86 | 132.37 | 315.61 | 0.192 |

### *Elastic constants*

The elastic properties of $Ti_2AlN$ and $Cr_2AlC$ were calculated[115] by using pseudopotential plane-wave methods and density functional theory. The exchange-correlation energy was deduced by means of generalized gradient approximation, and the elastic constants (table 35) were obtained by using the static finite strain technique. The bulk and shear moduli, Young's modulus and Poisson ratio for ideal polycrystalline aggregates were then derived (table 36). The difference between $C_{11}$ and $C_{33}$ was relatively small. The deviation of the isotropy factor from unity was larger for $Ti_2AlN$ than for $Cr_2AlC$; as intimated by the band structure.

*Table 37. Calculated elastic constants (GPa) and hardness (GPa) of $Ti_2AlN_x$*

| Sample | $C_{11}$ | $C_{12}$ | $C_{13}$ | $C_{33}$ | $C_{44}$ | G | B | E | H |
|---|---|---|---|---|---|---|---|---|---|
| $Ti_2AlN_{0.88}$ | 277.1 | 71.8 | 84.7 | 277.1 | 119.1 | 107.1 | 145.9 | 258.2 | 18.4 |
| $Ti_2AlN_{0.91}$ | 272.1 | 74.0 | 84.9 | 283.6 | 121.1 | 107.8 | 146.1 | 259.7 | 18.6 |
| $Ti_2AlN_{0.94}$ | 306.9 | 72.6 | 85.1 | 295.3 | 126.7 | 117.1 | 154.9 | 280.6 | 20.3 |
| $Ti_2AlN$ | 310.7 | 68.9 | 88.3 | 286.8 | 127.7 | 119.4 | 155.4 | 285.2 | 21.1 |

Powder metallurgy samples were prepared[116] with stoichiometric, hypo-stoichiometric or hyper-stoichiometric nitrogen contents. The presence of nitrogen vacancies resulted in a lattice contraction which occurred mainly along the c-axis. The elastic moduli and intrinsic hardness values of hypo-stoichiometric $Ti_2AlN_{0.9}$, as measured using nano-indentation tests, were slightly smaller than those for $Ti_2AlN$. The bulk and shear moduli and the Young's moduli, together with the hardness were calculated (table 37) using density functional theory. These exhibited differing responses, depending upon the concentration of nitrogen vacancies.

Titanium and AlN powders were prepared[117] in a 2:1 molar ratio and were subjected to high-energy milling under argon for 10h. The $Ti_2AlN$ was then prepared by the solid-state reaction of unmilled or milled powders during hot-pressing under 15 or 30MPa at 1200C for 2h (table 38). The microstructures and properties of the samples were related to the load which was used for hot-pressing. A marked increase in hardness and densification, and a smaller grain-size, were found for samples which were prepared from activated powders and were attributed to the formation of secondary phases such as $Ti_5Si_3$ and $Al_2O_3$.

*Table 38. Properties of mechanically-activated and non mechanically-activated $Ti_2AlN$*

| Condition | Hot-Pressing | Hardness (GPa) | Fracture Toughness (MPam$^{0.5}$) |
|---|---|---|---|
| Non-Activated | 1200C, 2h, 15MPa | 3.2 | 2.4 |
| Non-Activated | 1200C, 2h, 30MPa | 4.3 | 4.8 |
| Activated | 1200C, 2h, 15MPa | 5.6 | 3.7 |
| Activated | 1200C, 2h, 30MPa | 5.9 | 3.9 |

Amorphous and polycrystalline thin films were deposited[118] from a $Ti_2AlN$ target by means of magnetron sputtering, and the structures were altered by varying the deposition parameters and the cathode power-supply. As compared with amorphous and polycrystalline Ti-Al-N thin films, $Ti_2AlN$ film exhibited a greater hardness and toughness due to its nanolaminate structure. There was also a better wear resistance in ball-on-disk and nanowear tests. Multi-cycle measurements, with indentation depths of up to 100nm were performed, showing that the hardness, H, and elastic modulus, E, of amorphous film were 15.84 and 186.96GPa, respectively. These values for polycrystalline face-centred cubic film increased to 24.65 and 284.14GPa, respectively. The $Ti_2AlN$ film exhibited the highest (26.18GPa) hardness, but the elastic modulus (243.14GPa) was about 14% lower than that of the polycrystalline film. The H/E ratio was related to the elastic strain-to-failure, and similar H/E values were found for the amorphous (0.085) and polycrystalline (0.087) films, whereas a much higher H/E value (0.108) was found for $Ti_2AlN$ film. Polycrystalline $Ti_2AlN$ films were prepared[119] via the post-deposition annealing of Ti-Al-N film at 600 to 800C under high vacuum. Following post-deposition annealing at above 600C, the as-deposited amorphous Ti-Al-N film transformed into polycrystalline $Ti_2AlN$ film. Upon increasing the annealing temperature from 600 to 700C, the crystallinity of the polycrystalline $Ti_2AlN$ film was improved. The polycrystalline $Ti_2AlN$ film exhibited the greatest (34.1GPa) hardness while the hardness of amorphous Ti-Al-N films was only 24.2GPa. Thin films of $Ti_2AlN$ were deposited[120] onto (111) MgO substrates at between 500 and 750C by means of direct-current reactive magnetron sputtering of a $Ti_2Al$ compound target in a $N_2$/Ar plasma. The film initially consisted of a mixture of titanium, aluminium and (Ti,Al)N cubic solid solution at 500C, and changed into polycrystalline $Ti_2AlN$ at 600C. The crystallinity improved further with increasing substrate temperature. At 750C, a monocrystalline $Ti_2AlN$ (00•2) thin film formed which exhibited a characteristic layered hexagonal surface morphology, a high hardness, a high Young's modulus and a low electrical resistivity. The hardness increased from 15.8GPa at 600C to 26.5GPa at 750C, although the Young's modulus remained essentially constant at 412.5GPa. The $Ti_2AlN$ was stable at up to 600C. At 700C, aluminium preferentially desorbed from the surface, and the hardness and Young's modulus deteriorated[121]. Density functional theory calculations showed that aluminium atoms preferred to diffuse out horizontally from the $Ti_2AlN$ along the aluminium basal planes. When monocrystalline $Ti_2AlN$ (00•1) thin films were grown onto (111)-oriented MgO substrates at 830C, by using ultra-high vacuum direct-current reactive magnetron sputtering[122], there was found to be a narrow region for the growth of $Ti_2AlN$, with respect to the nitrogen content in the discharge. Perovskite $Ti_3AlN$ and intermetallic $Ti_3Al$ and TiAl phases predominated under low-nitrogen conditions. Under hyper-stoichiometric deposition conditions with respect to $Ti_2AlN$, a phase mixture which included NaCl-structured TiN was produced. Layer-by-layer epitaxial growth of all of the phases on the (00•1) basal planes was observed. Nano-indentation showed that the film hardness increased from 11 to 27GPa with increasing nitrogen content, with a corresponding phase transformation from Ti-Al intermetallics to $Ti_3AlN$, $Ti_2AlN$ and TiN.

When bulk samples of $Ti_2AlN$ with a density close to theoretical were synthesized[123] by spark plasma sintering, this yielded an homogenous distribution of grains. The microhardness values were essentially constant under loads of between 6 and 6.5GPa. Samples of $Ti_2AlN$-based material were prepared[124] via the mechanical activation of a Ti/AlN powder mixture in a planetary ball mill, followed by vacuum spark plasma sintering. The AlN/Ti ratio gradually decreased during mechanical activation. The maximum (90wt%) content of $Ti_2AlN$ was achieved by using a spark plasma sintering temperature of 1300C, and the lowest (1.9%) porosity was achieved by using a spark plasma sintering temperature of between 1200 and 1300C. The Vickers (0.5N) hardness was close to 7GPa. Monocrystalline (00•1) thin films of pure $Ti_2AlN$ were prepared[125] via ultra-high vacuum direct-current reactive magnetron sputtering onto (111)-oriented MgO substrates. The hexagonal-lattice parameters were c = 13.6Å and a = 3.07Å. The hardness and Young's modulus, as determined by nano-indentation, were 16.1 and 270GPa, respectively.

## Ti₄AlN₃

### Hardness

Dense bulk samples were prepared[126] by annealing (1300C, 4h), giving well-developed plate-like elongated grains with diameters of 2 to 5µm and thicknesses of 10 to 15µm. The Vickers hardness, fracture toughness and flexural strength were 2.7GPa, 350MPa, and 6.2MPam$^{0.5}$, respectively. The structures of the layered hexagonal nanolaminates, $Ti_3SiC_2$, $Ti_3AlC_2$ and $Ti_4AlN_3$, were analyzed[127] following preparation by reaction sintering to give a porosity of 3 to 5% and grain sizes of 3 to 15µm. Micro-indentation (0 to 1N) at room temperature and macro-indentation (10N) at 20 to 1200C revealed the modulus of elasticity, hardness and residual deformation. In accord with increasing hardness, the resistance to deformation and creep at intermediate and high temperatures of the materials were in the order: $Ti_3AlC_2$ - $Ti_4AlN_3$ - $Ti_3SiC_2$. At above 1000C, the indentation behaviours of $Ti_3AlC_2$ and $Ti_4AlN_3$ were close to one another, and very different to that of $Ti_3SiC_2$. The differences in the high-temperature properties were attributed to differences in the bond energies of aluminium atoms with each other and with titanium atoms, and to differing proximities of the testing temperature to the decomposition temperature. Bulk samples of $Ti_4AlN_3$ were fabricated[128] by reactive hot isostatic pressing (1275C, 24h, 70MPa) of $TiH_2$, AlN and TiN. Further annealing (1325C, 168h, argon) produced dense largely single-phase samples with less than 1vol% of TiN as a secondary phase. The machinable low-density (4.6g/cm$^3$) material with an average grain-size of 20µm was relatively soft, with a Vickers hardness of 2.5GPa. The Young's and shear moduli were 310 and 127GPa, respectively, while the corresponding compressive and flexural strengths at room temperature were 475 and 350MPa. At 1000C, the deformation was plastic, with a maximum compressive stress of about 450MPa. The greatest (50%) strength loss was found for a quenching temperature of 1000C. Further increases in the quenching temperature increased the retained strength before it finally decreased again. The material was damage-tolerant in that a 100N-load diamond indentation that produced a 0.4mm defect reduced the flexural strength by only 12%.

## Ti₃(Al,Si)C₂

### Hardness

High-purity $Ti_3Al_{1-x}Si_xC_2$ solid solutions, where x ranged from 0 to 1, were reaction-sintered[129] from titanium, silicon, aluminium and TiC powders using pulsed electric current sintering. The a-lattice parameter of the sintered solid solutions remained constant at about 0.307nm, while the c-lattice parameter decreased from 1.858 to 1.763nm, with increasing silicon content. The coefficient of thermal expansion was lower than those of the end members. The Young's and shear moduli increased with increasing silicon content. The Vickers hardness exhibited a marked hardening effect, regardless of the grain size; changing from 4.1GPa for $Ti_3AlC_2$ and 4.2GPa for $Ti_3SiC_2$, to 5.6GPa for $Ti_3(Al_{0.4}Si_{0.6})C_2$. The room-temperature strengthening effect was modest in fine-grained (7 x 3µm) samples, as the compressive strengths of $Ti_3Al_{0.6}Si_{0.4}C_2$ and $Ti_3Al_{0.4}Si_{0.6}C_2$ were higher by only 7.6% when compared with those of the end-members. A marked strengthening effect was observed in coarse-grained (25 x 8µm) samples however because the room-temperature compressive strengths of the solid solutions exceeded those of the end-members by more than 30%. When above the brittle-ductile transition temperature, the solid-solution strengthening effect decreased and the strength of $Ti_3SiC_2$ was much higher than that of $Ti_3AlC_2$ or the solid solutions.

## $Ti_3(Al,Si,Sn)C_2$

### *Hardness*

Almost pure solid-solution powders were prepared[130] using titanium, aluminium, silicon, tin and TiC powders as raw materials and pressureless sintering (1450C, 600s, Ar atmosphere). The $Ti_3Al_{0.8}Si_{0.2}Sn_{0.2}C_2$ grains had a typical lamellar structure. Essentially fully dense bulk $Ti_3Al_{0.8}Si_{0.2}Sn_{0.2}C_2$ was prepared by two-step hot-pressing (1450C, 0.5h, 30MPa), and the flexural strength and Vickers hardness (under a load of 4.9 N), were 649MPa and 6.4GPa, respectively. These values were higher than those for single-phase $Ti_3AlC_2$, or for $Ti_3AlSn_{0.2}C_2$ and $Ti_3AlSi_{0.2}C_2$ solid solutions; reflecting solid-solution strengthening. Grain delamination and kink-band formation was extensive in indentations and on fracture surfaces.

## $Ti_2(Al,Sn)C$

### *Hardness*

High-purity $Ti_2Al_{1-x}Sn_xC$ powders, where x ranged from 0 to 1, were prepared[131] by pressureless sintering using titanium, aluminium, tin and TiC as raw materials. Pure $Ti_2AlC$ and $Ti_2Al_{0.8}Sn_{0.2}C$ were obtained when the above raw materials were in the ratios of Ti:1.1Al:0.9TiC and Ti:0.9Al:0.2Sn:0.9TiC, respectively, at 1450C. Fully dense bulk specimens of $Ti_2AlC$ and $Ti_2Al_{0.8}Sn_{0.2}C$ were produced by mechanical alloying and hot-pressing and sintering. The *a* lattice parameter increased, while the *c* lattice parameter remained essentially constant, as the tin content was increased. The Vickers hardness of

Materials Research Forum LLC
https://doi.org/10.21741/9781644901274

Ti$_2$AlC and Ti$_2$Al$_{0.8}$Sn$_{0.2}$C approached 6 and 4GPa, respectively, while the corresponding flexural strengths were 650 and 521MPa. Grain delamination, kink bands and crack deflection occurred around indentations and at fracture surfaces.

*Table 39. Comparison of the elastic constants (GPa) of various MAX phases*

| Phase | C$_{11}$ | C$_{33}$ | C$_{44}$ | C$_{66}$ | C$_{12}$ | C$_{13}$ |
|---|---|---|---|---|---|---|
| Ti$_2$AsC | 224.0 | 293.9 | 140.6 | 33.9 | 156.9 | 123.9 |
| Ti$_3$AlC$_2$ | 358.1 | 292.6 | 122.0 | 136.0 | 83.9 | 74.8 |
| Ti$_3$SiC$_2$ | 370.0 | 349.4 | 150.6 | 135.4 | 99.2 | 110.6 |
| Ti$_3$GeC$_2$ | 356.8 | 324.7 | 128.1 | 127.1 | 99.6 | 97.0 |
| Ti$_2$AlC | 300.5 | 266.4 | 106.4 | 115.8 | 68.9 | 61.7 |
| Ti$_2$GaC | 303.2 | 254.6 | 99.1 | 116.8 | 72.1 | 62.6 |
| Ti$_2$InC | 290.5 | 236.3 | 86.4 | 112.3 | 70.7 | 58.0 |
| Ti$_2$SiC | 311.4 | 324.2 | 146.1 | 112.5 | 85.8 | 111.5 |
| Ti$_2$GeC | 286.3 | 288.9 | 119.8 | 99.7 | 84.1 | 99.9 |
| Ti$_2$SnC | 252.8 | 254.4 | 92.9 | 79.2 | 91.4 | 74.1 |
| Ti$_2$PC | 269.2 | 344.3 | 173.8 | 66.1 | 132.3 | 151.3 |
| Ti$_2$SC | 339.3 | 359.9 | 161.2 | 116.8 | 101.4 | 110.1 |
| Ti$_2$AlN | 316.0 | 290.7 | 128.6 | 123.1 | 72.0 | 94.8 |
| V$_2$AlC | 330.1 | 316.9 | 147.9 | 127.9 | 73.0 | 104.3 |
| Nb$_2$AlC | 314.7 | 295.4 | 138.6 | 112.3 | 88.8 | 117.3 |
| Cr$_2$AlC | 364.5 | 356.1 | 139.8 | 140.0 | 84.4 | 107.4 |
| Ta$_2$AlC | 348.5 | 337.8 | 154.7 | 115.4 | 118.9 | 132.7 |
| a-Ta$_3$AlC$_2$ | 441.2 | 381.7 | 174.8 | 154.7 | 132.1 | 138.0 |
| a-Ta$_4$AlC$_3$ | 445.0 | 378.2 | 178.7 | 141.3 | 162.6 | 150.8 |
| Ta$_5$AlC$_4$ | 466.5 | 410.4 | 187.6 | 158.6 | 147.6 | 164.4 |

*Table 40. Comparison of the elastic moduli (GPa) of various MAX phases*

| Phase | K | G | E | Poisson Ratio |
|---|---|---|---|---|
| $Ti_2AsC$ | 166.7 | 71.7 | 188.2 | 0.312 |
| $Ti_3AlC_2$ | 163.1 | 127.3 | 303.1 | 0.190 |
| $Ti_3SiC_2$ | 192.2 | 138.1 | 334.3 | 0.210 |
| $Ti_3GeC_2$ | 180.4 | 126.2 | 307.1 | 0.216 |
| $Ti_2AlC$ | 138.8 | 110.6 | 262.2 | 0.185 |
| $Ti_2GaC$ | 138.9 | 106.9 | 255.2 | 0.194 |
| $Ti_2InC$ | 131.5 | 98.3 | 236.1 | 0.201 |
| $Ti_2SiC$ | 173.6 | 122.0 | 296.6 | 0.215 |
| $Ti_2GeC$ | 158.7 | 105.8 | 259.6 | 0.227 |
| $Ti_2SnC$ | 137.6 | 87.5 | 216.5 | 0.238 |
| $Ti_2PC$ | 191.5 | 102.7 | 261.3 | 0.273 |
| $Ti_2SC$ | 186.7 | 134.1 | 324.6 | 0.210 |
| $Ti_2AlN$ | 160.6 | 119.6 | 287.4 | 0.202 |
| $V_2AlC$ | 171.1 | 130.0 | 311.1 | 0.197 |
| $Nb_2AlC$ | 174.6 | 116.3 | 285.6 | 0.227 |
| $Cr_2AlC$ | 187.0 | 136.1 | 328.6 | 0.207 |
| $Ta_2AlC$ | 200.3 | 126.5 | 313.5 | 0.239 |
| $a-Ta_3AlC_2$ | 230.8 | 157.0 | 383.9 | 0.223 |
| $a-Ta_4AlC_3$ | 243.2 | 151.7 | 376.8 | 0.242 |
| $Ta_5AlC_4$ | 466.5 | 163.0 | 403.1 | 0.236 |

## Ti$_2$AsC

### *Elastic constants*

The intrinsic mechanical properties of this, and of various other MAX phases (tables 39 and 40) were calculated[132] by using first-principles density functional theory. A stress-versus-strain approach was used to predict the bulk modulus, shear modulus, Young's modulus and Poisson's ratio; based upon the Voigt-Reuss-Hill approximation for polycrystals. These predictions were in good agreement with experimental data. There was an inverse relationship between the Poisson and Pugh ratios of the shear and bulk moduli. A higher Poisson ratio generally implied a higher fracture energy. The calculations also predicted that Ti$_2$AsC and Ti$_2$PC should have a much higher ductility than those of other such compounds. It was concluded that MAX phases exhibit a wide range of properties, ranging from very ductile to brittle. The nature of the A-component of the MAX phase was found to be the most important factor. The measured Vickers hardness of the compounds did not appear to be related to any of the mechanical parameters. It was suggested to be possible to predict new MAX phases that are less brittle or more tough by engineering a higher G/K ratio.

*Table 41. Unit cell parameters of Ti$_2$GeC as a function of pressure*

| Pressure (GPa) | a (Å) | c (Å) |
|---|---|---|
| 0 | 3.078 | 12.934 |
| 3.35 | 3.066 | 12.856 |
| 5.72 | 3.056 | 12.799 |
| 11.29 | 3.032 | 12.680 |
| 17.02 | 3.014 | 12.571 |
| 30.04 | 2.976 | 12.379 |
| 39.73 | 2.945 | 12.239 |
| 49.47 | 2.922 | 12.135 |

## $Ti_2GeC$

### *Elastic Constants*

The layered material with space group, $P6_3/mmc$, *a* lattice parameter of 3.078Å and *c* lattice parameter of 12.933Å, was prepared[133] by hot-pressing and the pressure dependences of the lattice parameters were measured. The bulk modulus, as calculated using the Birch-Murnaghan equation of state, was 211GPa. This value was greater, than those for $Ti_2AlC$ and $Ti_2SC$, by 13 and 10%, respectively. It was also the highest value among those for $Ti_2GeC$, $V_2GeC$ and $Cr_2GeC$. The room-temperature density was 5.48g/cm3; 97% of the theoretical density. The pressure dependence of the lattice parameters indicated that the phase was structurally stable at least up to 49.47GPa (table 41).

## $Ti_3GeC_2$

### *Hardness*

Polycrystalline fully-dense largely single-phase samples of $Ti_3Si_{0.5}Ge_{0.5}C_2$, $Ti_3Si_{0.75}Ge_{0.25}C_2$, and $Ti_3GeC_2$ (a = 3.07Å, c = 17.76Å, theoretical density = 5.55g/cm$^3$) having various grain-sizes were prepared[134] by reactive hot isostatic pressing or hot-pressing. It was found that solid solubility in $Ti_3(Si_xGe_{1-x})C_2$ ranged from x = 0 to at least x = 0.75. The hardness (2.5GPa) of both solid-solution compositions was between those of $Ti_3SiC_2$ (3.0GPa) and $Ti_3GeC_2$ (2.2GPa), and it was thus concluded that no solid-solution strengthening occurred in this system. All of the samples were damage-tolerant and thermal-shock resistant. A 300N Vickers indentation, made in a 1.5mm-thick 4-point bend bar, decreased its strength by between 25 and 35%. Quenching into water from 1000C reduced the 4-point flexural strength by 10 to 20%. The post-quenching flexural strength of the coarse-grained $Ti_3Si_{0.5}Ge_{0.5}C_2$ samples was about 25% higher than that of as-received bars. Increasing the germanium content decreased the compressive strength. The ultimate compressive strength of fine-grained $Ti_3Si_{0.5}Ge_{0.5}C_2$ decreased monotonically from room temperature to 950C. The failure was brittle at room temperature but, at above 1000C, the loss in strength was greater but deformation was more plastic. Bulk $Ti_3(Al_{1-x}Ge_x)C_2$ samples, where x was 0, 0.3, 0.5, 0.7 or 1.0, were prepared[135] by the reactive hot-pressing (1400C, 1h, 25MPa) of $TiC_{0.6}$, aluminium and germanium powders. With increasing aluminium or germanium content, the Vickers hardness and flexural strength increased relative to the properties of the end-members. The maximum flexural strength of $Ti_3(Al_{0.7}Ge_{0.3})C_2$ was about 600MPa; about 3 times higher than that of pure $Ti_3GeC_2$. The Vickers hardness values decreased slightly with increasing indentation load.

### Elastic constants

The constants, $C_{11}$, $C_{12}$, $C_{13}$, $C_{33}$ and $C_{44}$, have been calculated[136] to be 355, 143, 80, 404 and 172GPa, respectively. The Young's and shear moduli were 345 and 144GPa, respectively.

## $Ti_2InB_2$

### Hardness

The properties of this new 212-type MAB phase were compared[137] with those of $Ti_2InC$, $Ti_2SnC$ and $Ti_2AlC$. In the case of $Ti_2InB_2$, B-B bonds existed in higher numbers, indicating a highly covalent nature which made it stiffer than phases such as $Ti_2InC$. The $Ti_2InB_2$ therefore exhibited high $C_{11}$, $C_{33}$, B, G and E values when compared with $Ti_2InC$. The calculated hardness of $Ti_2InB_2$ was the highest among all of the present phases.

## $(Ti,Mo)_2AlC$

### Hardness

The creation of $(Ti_{0.9}Mo_{0.1})_2AlC$ by reacting titanium, aluminium, TiC and molybdenum powder mixtures was investigated[138] at 500 to 1450C. This showed that reaction between titanium and aluminium produced Ti-Al intermetallics and reaction between aluminium and molybdenum produced Mo-Al intermetallics. The above phase was then formed by the reaction of Ti-Al and Mo-Al intermetallics and TiC. The pure $(Ti_{0.9}Mo_{0.1})_2AlC$ phase had a Vickers hardness, flexural strength and fracture toughness of 5.48GPa, 363.60MPa and $5.78MPam^{0.5}$, respectively. The density increased from 4.09 to $4.32g/m^3$, and the Vickers hardness increased from 3.79 to 5.47GPa when the molybdenum content was increased from 0 to 20at%. A similar strengthening effect of molybdenum additions was observed in the flexural strength and fracture toughness. The flexural strength increased smoothly from 270.58 to 363.60MPa when the molybdenum content was increased from 0 to 20at%.

### Toughness

The $K_{Ic}$ value first increased, and then flattened out when the molybdenum content exceeded 10at%. It increased from 5.48 to $5.77MPam^{1/2}$ as the molybdenum content increased from 10 to 20at%. As compared with monolithic $Ti_2AlC$, the flexural strength and fracture toughness were increased by 34 and 136%, respectively, in the case of $(Ti_{0.8}Mo_{0.2})_2AlC$. The material was strengthened by replacing titanium with molybdenum because the latter has a smaller atomic size and one more valence electron.

## $Ti_2SC$

### *Hardness*

High-purity material (a = 3.21Å, c = 11.22Å) was prepared[139] by using a simple two-step spark plasma sintering technique which required a lower (1450C) sintering temperature and lower (30MPa) pressure. The purity and microstructure could be controlled by varying the molar ratio of the $TiS_2$, titanium and carbon precursors. The material had a low (4.60g/cm$^3$) density, with a Vickers hardness of 6.7GPa, a fracture toughness of 5.4MPam$^{0.5}$ and a flexural strength of 394MPa. First-principles studies of the elastic properties showed[140] that the absence of a band-gap at the Fermi level, and a finite value of the density-of-states at the Fermi energy, revealed the metallic nature of the material. The 5 independent elastic constants were deduced, and the bulk modulus, Young's modulus, shear modulus and Poisson ratio were determined, with $Ti_2SC$ being elastically stable and exhibiting a high elastic isotropy. A high bulk modulus and hardness were attributed to strong $Ti_{3d}$-$S_{2p}$ hybridization. The existence of such strong M-A bonding is unusual among MAX phases. This layered machinable material has a much smaller c-lattice parameter than that of other MAX phases and this was expected to improve the mechanical behaviour. Fine-grained (2 to 4μm) and coarse-grained (10 to 20μm) polycrystalline fully-dense samples were prepared[141] by hot-pressing (1500C, 5h, 45MPa) and further annealing (1600C, 20h). They comprised some 6vol% of an impurity anatase phase. The average Vickers hardness (2 to 300N) was 8GPa, and the fine-grained samples were slightly harder. Unlike other MAX phases, cracks extended here from the corners of the Vickers indentations in fine-grained specimens. On the basis of these cracks, the fracture toughness was estimated to increase essentially linearly, from 4 to 6MPam$^{0.5}$, as the Vickers load was increased from 50 to 300N. The room-temperature compressive strength of fine-grained samples was 1.4GPa, and the failure mode was brittle. The Young's modulus was 316GPa. There was no evidence of kink-band formation during compression and this was attributed to the fine grain-size. A first-principles study[142] of the phase stability and compression behavior of $Ti_2SC$ and $Ti_2AlC$ showed that the former was more stable, due to its deeper pseudo-gap, lower density-of-states at the Fermi level and stronger titanium-sulfur hybridization. The $Ti_2SC$ also had a higher stiffness when compressed, with the calculated bulk modulus being 181 and 162GPa for $Ti_2SC$ and $Ti_2AlC$, respectively. The bond strength of both materials became greater at high pressures, but there was no apparent change in the nature of the bonding under pressure.

### Elastic constants

This material tends to be an outlier in many respects, and this is true of the elastic constants. There are discrepancies in the values reported (table 42) when differing conditions are assumed.

*Table 42. Elastic constants of $Ti_2SC$*

| $C_{11}$ (GPa) | $C_{12}$ (GPa) | $C_{13}$ (GPa) | $C_{33}$ (GPa) | $C_{44}$ (GPa) | $C_{66}$ (GPa) | Reference |
|---|---|---|---|---|---|---|
| 368.0 | 108.0 | 123.0 | 395.0 | 189.0 | 130.0 | [143] |
| 319.0 | 101.0 | 106.0 | 351.0 | 149.0 | 109.0 | [144] |
| 339.0 | 90.0 | 100.0 | 354.0 | 162.0 | 125.0 | [145] |

### $Ti_3(Si,Al)C_2$

### Hardness

Samples of $Ti_3Si_{1-x}Al_xC_2$ were prepared[146], where x ranged from 0 to 1, via the spark plasma sintering of TiC, titanium, silicon and aluminium powder at 1200C. The impurities were TiC and $Ti_5Si_3$. The former was present in all of the samples, while trace amounts of $Ti_5Si_3$ were detected when x was between 0.1 and 0.6. The a- and c- lattice parameters increased linearly with increasing aluminium content across the entire x-range, from 3.0676 and 17.6823Å, respectively, for x = 0, to 3.0676 and 17.6823, respectively, for x = 1, and obeyed Vegard's law. The density and hardness (figure 34) decreased as the aluminium content increased.

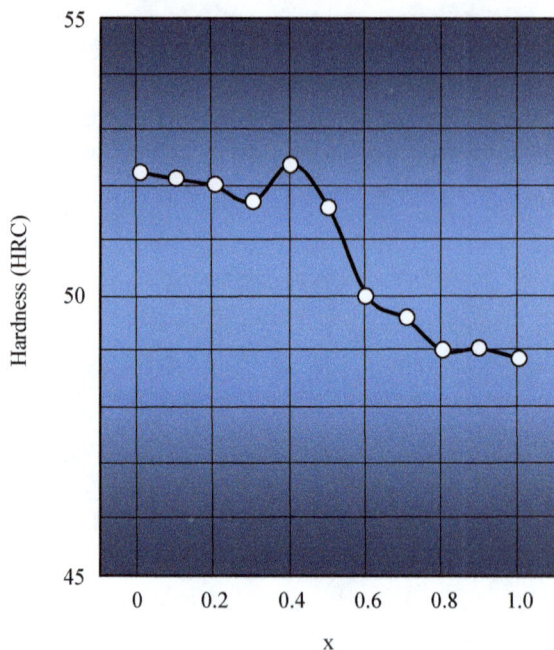

*Figure 34. Hardness of Ti₃Si₁₋ₓAlₓC₂ as a function of x*

## TiSiC

### *Hardness*

This phase belongs to the P6₃/mmc space group, with an *a* lattice parameter of 3.077Å and a *c* lattice parameter of 17.68Å. It has been found that microwave processing leads[147] to higher hardness values and compressive strengths, some 8GPa and 240MPa respectively, than does conventional sintering. The associated densities were 3.47 and 4.154g/cm³ for conventional heating and microwave treatment, respectively, while the theoretical density was 4.51g/cm³. The microhardness of 8.2GPa after microwave processing was about twice that (4.2GPa) of conventionally prepared samples. In the case of compressive strength, microwave sintering led to values which were some 3 times higher than those for conventional processing. The specific strengths were correspondingly affected, with microwave treatment yielding values ranging from 53000 to 63000m$^2$/s$^2$ while conventional methods yielded values ranging from 19000 to 25000m$^2$/s$^2$.

## Ti$_3$SiC$_2$

### *Hardness*

This material (a = 3.07Å, c = 17.67Å, theoretical density = 4.52g/cm$^3$) was one of the earliest MAX phases to be noticed because it was seen to be anomalously soft for a transition-metal carbide. The hardness was also reported to be anisotropic, with the values measured normal to the basal planes being some three times higher than the values which were measured parallel to the basal planes.

Bulk material has been synthesized[148] by reactive melt infiltration. Powders having various chemical compositions were pelleted by cold-pressing, and infiltration was performed at 1500C for 1h under 10$^{-4}$torr. A pre-form composition of 3TiC/0.3Si was found to be the optimum one. Adding small amounts of silicon to the pre-form composition improved the purity of the Ti$_3$SiC$_2$. The relative density, Vickers hardness and Young's modulus were 74.7%, 4.1GPa and 290.2GPa, respectively, for samples which contained the highest Ti$_3$SiC$_2$ content. The relative density, Vickers hardness and Young's modulus were 97.5%, 6.68GPa and 306.41GPa, respectively, for materials produced using the optimum milling time of 1h. These samples contained 84.8% of Ti$_3$SiC$_2$. The relative density, Vickers hardness and Young's modulus were 83.3%, 6.37GPa and 178.23GPa, respectively, for material produced using the optimum applied pressure of 40MPa. These samples contained 93% of Ti$_3$SiC$_2$. Those samples which had the optimum purity exhibited a relative density, Vickers hardness and Young's modulus of 83.3%, 6.4GPa and 178.2GPa, respectively. Upon increasing the amount of infiltrated silicon from 1 to 3 times the stoichiometric value, the fraction of Ti$_3$SiC$_2$ increased and attained its maximum value at 3 times the stoichiometric value.

The room-temperature abrasive wear behavior of Ti$_3$SiC$_2$, solution-strengthened Ti$_2.7$Zr$_0.3$SiC$_2$ and Cr$_2$AlC, was investigated[149] by means of low-velocity scratch-testing using a conical diamond indenter and loads of up to 20N. The wear-rate was essentially a linear function of the applied load. The Ti$_3$SiC$_2$, with a hardness of 2.8GPa, exhibited the lowest wear resistance with extensive plowing and grain-breakout damage, especially at the highest load. The hardest (7.3GPa) material was the Ti$_2.7$Zr$_0.3$SiC$_2$ and exhibited a 5 times better wear resistance. The Cr$_2$AlC, with a hardness of 4.8GPa, exhibited a wear resistance that was equal to, or better than, that of Ti$_2.7$Zr$_0.3$SiC$_2$. The wear mechanism depended upon the applied load and the microstructure of the material. In the case of

$Ti_3SiC_2$, quasi-plastic deformation occurred below a load of 10N and resulted in grain-bending, kink-band formation and delamination, grain de-cohesion and trans- and intra-granular fracture near to the scratch. At the same load, the $Ti_{2.7}Zr_{0.3}SiC_2$ and $Cr_2AlC$ samples suffered plastic plowing, grain-boundary cracking and material dislodgment.

Nano-layered $Ti_3SiC_2$ ceramics have been fabricated[150,151] via the liquid-silicon infiltration of gelcast porous TiC pre-forms. The $Ti_3SiC_2$ which was formed decomposed into TiC with increasing infiltration time. Samples with 88wt% of $Ti_3SiC_2$ were produced by infiltration at 1500C for 1h with a 50vol% solid loading and 10wt% monomer content. The hardness and flexural strength of the $Ti_3SiC_2$-based ceramic decreased with decreasing SiC and TiC impurity content, and attained 5.8GPa and 420MPa, respectively, for samples with 15wt% of impurity.

Microstructural and mechanical property changes in 7MeV $^{26}Xe^+$ ion-bombarded $Ti_3SiC_2$ and $Ti_3AlC_2$ have been monitored[152] at room temperature. The $Ti_3AlC_2$ exhibited a better resistance to irradiation damage, but exhibited a more marked transition from α- to β-phase. The fractions of β phase in $Ti_3SiC_2$ and $Ti_3AlC_2$ samples, bombarded at room temperature to a dose of $2 \times 10^{15}/cm^2$, were 28.05 and 57.42%, respectively. At a dose of $4 \times 10^{15}/cm^2$, the $Ti_3AlC_2$ lattice was barely damaged, while that of $Ti_3SiC_2$ was heavily distorted. The hardness of both materials was increased by room-temperature bombardment, due to defect-pinning.

Epitaxial (00•1) $Ti_3SiC_2$ thin films have been deposited[153] onto (111) MgO and (00•1) $Al_2O_3$ substrates at 800 to 900C by direct-current magnetron sputtering, using titanium, carbon and silicon targets. Deposition onto the MgO yielded phases with (10•5) as the preferred orientation. A low (25μcm) resistivity was found for as-deposited single-crystal $Ti_3SiC_2$ films. The Young's modulus, as measured by nano-indentation, was 343 to 370GPa. The mechanical deformation, using cube-corner and Berkovich indenters, indicated an initially high hardness of almost 30GPa. With increasing maximum indentation load, the hardness decreased towards bulk values. This was attributed to kink formation, with dislocation-ordering and delamination at basal planes. Samples of $Ti_4SiC_3$, $Ti_5Si_2C_3$ and $Ti_7Si_2C_5$ were also studied[154]. The latter two materials were regarded as being intergrowths. Epitaxial films of $Ti_5Si_3C_x$ could be deposited, and $Ti_5Si_4$ was also observed. The $Ti_5Si_2C_3$ and $Ti_7Si_2C_5$ were deemed to be metastable phases. The hardness was related to the number of silicon layers per titanium layer. Pulsed laser deposition could produce[155] good-quality thin films with a surface roughness of 0.46nm, a friction coefficient of 0.2 in humid air and a hardness of between 30 and 40GPa. An anisotropic layer structure of the $Ti_3SiC_2$, and nanocrystallites in coatings, were related to a low friction and high hardness. Ceramics which comprised $Ti_3SiC_2$ and $Ti_4SiC_3$ were prepared[156] via the high-temperature vacuum reduction of $TiO_2$ by SiC, followed by hot-pressing (1600C, 25MPa). The hot-pressing step did not greatly affect the composition, thus reflecting the good stability of the MAX phases. A lower ductility was found for $Ti_4SiC_3$ than for $Ti_3SiC_2$. The observed flexural strength and fracture toughness, and the microhardness of the softer parts (table 43), were typical of coarse-grained MAX-phase ceramics.

When this material is thinned by various means, the A-element layer can shear. Such a dimensionally-induced phase transformation, which involves shearing of the silicon planes, can also by shearing of the aluminium planes in $Ti_3AlC_2$.

*Table 43. Properties of $Ti_4SiC_3$-$Ti_3SiC_2$ hot-pressed ceramics*

| $Ti_3SiC_2$ | $Ti_4SiC_3$ | TiC | SiC (vol%) | Strength (MPa) | K (MPam$^{0.5}$) | H$_V$ (GPa) |
|---|---|---|---|---|---|---|
| 60 | - | 35 | 5 | 363 | 5.35 | 5.4 |
| 15 | 70 | 15 | - | 241 | 4.60 | 5.1 |
| - | 85 | 15 | - | 232 | 4.25 | 4.9 |

Most of the dislocations in $Ti_3SiC_2$ are basal dislocations, with a Burgers vector of $1/3<1\bar{2}\bullet0>$. Non-basal dislocations would have a Burgers vector that was greater than the **c** lattice parameter. The dislocations tend to be mixed, with both edge and screw components. The dislocations prefer to arrange themselves into walls of low-angle or high-angle grain boundaries that are normal to the basal planes. They can also form pile-ups on those basal planes. The walls possess both tilt and twist components. Low-angle kink boundaries are also mobile. Boundaries can be viewed as comprising alternating parallel mixed perfect dislocations, having differing Burgers vectors in the basal plane with a relative angle of 120°. A preponderance of one type of dislocation can then account for twisting. An overall result of the dislocation behaviour is that work-hardening tends not to occur. The critical resolved shear stress of the basal dislocations also tends to be very low.

*Table 44. Calculated elastic constants of undoped materials
and of materials doped at the titanium site*

| Phase | Dopant | $C_{11}$ (GPa) | $C_{33}$ (GPa) | $C_{44}$ (GPa) | $C_{12}$ (GPa) | $C_{13}$ (GPa) |
|-------|--------|------|------|------|------|------|
| $Ti_3AlC_2$ | - | 355.2 | 292.4 | 119.3 | 82.2 | 76.6 |
| $Ti_3AlC_2$ | Hf | 348.0 | 290.9 | 115.0 | 88.5 | 75.3 |
| $Ti_3AlC_2$ | Nb | 353.0 | 295.3 | 120.4 | 86.6 | 75.7 |
| $Ti_3AlC_2$ | Zr | 343.8 | 287.8 | 113.9 | 88.4 | 74.6 |
| $Ti_3SiC_2$ | - | 364.2 | 351.8 | 157.2 | 95.4 | 112.5 |
| $Ti_3SiC_2$ | Hf | 362.2 | 354.2 | 142.9 | 97.8 | 107.3 |
| $Ti_3SiC_2$ | Nb | 366.0 | 355.8 | 150.2 | 96.0 | 110.3 |
| $Ti_3SiC_2$ | Zr | 359.3 | 351.4 | 142.3 | 97.5 | 107.0 |

### Elastic constants

It was noted early on that this material is relatively stiff elastically, with Young's and shear moduli of the order of 330 and 133GPa, respectively. It is nevertheless machinable. The relative softness, combined with a high stiffness/hardness ratio, recalled ductile metals rather than ceramics. This leads to the common characterization of MAX phases as being so-called 'ductile ceramics'.

The specific stiffness of $Ti_3SiC_2$ is high because of its relatively low density. The specific stiffness is thus some three times higher than that of titanium. It is also comparable to that of a ceramic such as silicon nitride, but offers the advantage of being machinable.

*Table 45. Calculated moduli of undoped materials
and of materials doped at the titanium site*

| Phase | Dopant | B (GPa) | G (GPa) | E (GPa) | Poisson Ratio |
|-------|--------|---------|---------|---------|---------------|
| $Ti_3AlC_2$ | - | 163.0 | 125.8 | 300.2 | 0.2 |
| $Ti_3AlC_2$ | Hf | 162.0 | 121.5 | 291.6 | 0.2 |
| $Ti_3AlC_2$ | Nb | 163.4 | 125.4 | 299.6 | 0.2 |
| $Ti_3AlC_2$ | Zr | 160.4 | 120.0 | 288.2 | 0.2 |
| $Ti_3SiC_2$ | - | 191.2 | 139.6 | 336.9 | 0.2 |
| $Ti_3SiC_2$ | Hf | 189.3 | 134.5 | 326.2 | 0.2 |
| $Ti_3SiC_2$ | Nb | 191.2 | 138.1 | 333.9 | 0.2 |
| $Ti_3SiC_2$ | Zr | 188.1 | 133.4 | 323.8 | 0.2 |

The effects of doping with hafnium, niobium or zirconium upon the structural and mechanical properties of $Ti_3AlC_2$ and $Ti_3SiC_2$ were deduced[157] on the basis of density functional theory. Calculations showed that magnetism could be induced in $Ti_3AlC_2$ by doping certain sites with hafnium or zirconium, due to electron-transfer. Doping with any of the above dopants at one type of site had a minimal effect upon the mechanical properties of either material, while doping at a different site impaired either material; reducing the bulk, shear and Young's moduli. The structures consisted of 2 edge-shared $Ti_6C$ octahedral layers which were linked by a 2-dimensional close-packed aluminium or silicon plane where 2 different titanium atoms occupied the 2a point. This was termed a Ti1 site. The calculated elastic constants (table 44) and moduli (table 45) of materials which were doped at that site were comparable to those of undoped samples. Similar doping at an interstitial site markedly reduced the moduli (table 46), with the decrease in the shear and Young's moduli being greater than that of the bulk modulus. The reduced shear and Young's moduli were closely related to large decreases in the $C_{33}$ and $C_{44}$ elastic constants (table 47). Both phases had highly directional covalent bond-chains which were linked by weakly bonded aluminium or silicon atoms so as to form layered structures. The moduli were mainly due to strong Ti-C bonding, while easy basal-plane slip was attributed to the above weak bonding. Samples which were doped at the Ti1 site retained the bonding of the undoped materials.

*Table 46. Calculated moduli of undoped materials*
*and of materials doped at an interstitial site*

| Phase | Dopant | B (GPa) | G (GPa) | E (GPa) | Poisson Ratio |
|-------|--------|---------|---------|---------|---------------|
| Ti$_3$AlC$_2$ | - | 163.0 | 125.8 | 300.2 | 0.2 |
| Ti$_3$AlC$_2$ | Hf | 141.6 | 88.7 | 220.2 | 0.2 |
| Ti$_3$AlC$_2$ | Nb | 150.9 | 104.5 | 254.7 | 0.2 |
| Ti$_3$AlC$_2$ | Zr | 139.0 | 89.5 | 221.1 | 0.2 |
| Ti$_3$SiC$_2$ | - | 191.2 | 139.6 | 336.9 | 0.2 |
| Ti$_3$SiC$_2$ | Hf | 169.2 | 93.2 | 236.3 | 0.3 |
| Ti$_3$SiC$_2$ | Nb | 176.8 | 102.8 | 258.2 | 0.3 |
| Ti$_3$SiC$_2$ | Zr | 167.5 | 93.5 | 236.6 | 0.3 |

*Table 47. Calculated elastic constants of undoped materials*
*and of materials doped at an interstitial site*

| Phase | Dopant | C$_{11}$ (GPa) | C$_{33}$ (GPa) | C$_{44}$ (GPa) | C$_{12}$ (GPa) | C$_{13}$ (GPa) |
|-------|--------|---------|---------|---------|---------|---------|
| Ti$_3$AlC$_2$ | - | 355.2 | 292.4 | 119.3 | 82.2 | 76.6 |
| Ti$_3$AlC$_2$ | Hf | 310.7 | 176.7 | 108.6 | 91.6 | 85.1 |
| Ti$_3$AlC$_2$ | Nb | 326.3 | 226.2 | 116.0 | 91.7 | 83.4 |
| Ti$_3$AlC$_2$ | Zr | 310.5 | 181.6 | 110.9 | 87.8 | 78.8 |
| Ti$_3$SiC$_2$ | - | 364.2 | 351.8 | 157.2 | 95.4 | 112.5 |
| Ti$_3$SiC$_2$ | Hf | 318.3 | 238.0 | 99.1 | 119.7 | 106.8 |
| Ti$_3$SiC$_2$ | Nb | 328.5 | 276.5 | 106.4 | 118.1 | 111.6 |
| Ti$_3$SiC$_2$ | Zr | 314.0 | 239.4 | 97.1 | 120.6 | 103.9 |

*Table 48. Elastic constants (GPa) of Ti$_2$AN phases*

| Phase | C$_{11}$ | C$_{12}$ | C$_{13}$ | C$_{33}$ | C$_{44}$ |
|---|---|---|---|---|---|
| Ti$_2$SiN | 280.40 | 110.72 | 128.94 | 347.22 | 155.36 |
| Ti$_2$GeN | 230.13 | 117.87 | 124.34 | 283.46 | 119.73 |
| Ti$_2$SnN | 220.29 | 85.3 | 80.93 | 251.45 | 78.31 |

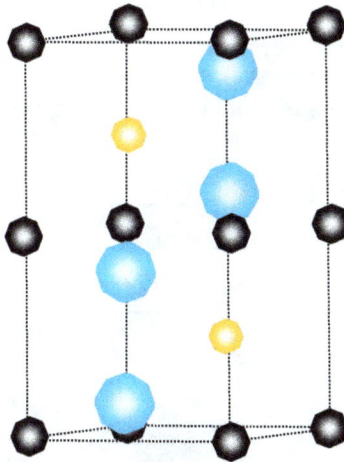

| A | a (Å) | c (Å) | c/a |
|---|---|---|---|
| Si | 2.992 | 12.779 | 4.27 |
| Ge | 3.046 | 12.907 | 4.24 |
| Sn | 3.176 | 13.468 | 4.24 |

*Figure 35. Crystal structure of Ti$_2$AN phases.*
*Black: nitrogen, blue: titanium, yellow: silicon, germanium or tin*

### Ti₂SiN

*Hardness*

First-principles density functional theory calculations within the generalized gradient approximation were used[158] to determine the structural (figure 35) and mechanical properties of $Ti_2AN$ phases, where A was silicon, germanium or tin. The elastic constants (table 48) were estimated by using the stress-strain method while the hardnesses and the isotropic elastic moduli (table 49) were studied within the Voigt–Reuss–Hill approximation for ideal polycrystalline aggregates.

*Table 49. Properties of Ti₂AN phases*

| Phase | B (GPa) | G (GPa) | E (GPa) | Poisson Ratio | Hᵥ (GPa) |
|-------|---------|---------|---------|---------------|----------|
| Ti₂SiN | 181.39 | 110.57 | 275.69 | 0.247 | 14.58 |
| Ti₂GeN | 163.14 | 79.52 | 205.21 | 0.290 | 8.16 |
| Ti₂SnN | 131.65 | 74.24 | 187.47 | 0.263 | 9.71 |

*Table 50. Lattice parameters (Å) of Ti₃(SnₓAl₁₋ₓ)C₂*

| x | a | c | c/a |
|------|-------|--------|-------|
| 0 | 3.080 | 18.642 | 6.053 |
| 0.25 | 3.095 | 18.666 | 6.031 |
| 0.5 | 3.115 | 18.681 | 5.997 |
| 0.75 | 3.131 | 18.684 | 5.967 |
| 1 | 3.151 | 18.698 | 5.934 |

*Table 51. Elastic constants (GPa) of Ti₃(SnₓAl₁₋ₓ)C₂*

| x | C₁₁ | C₃₃ | C₄₄ | C₁₂ | C₁₃ |
|------|-------|-------|-------|------|------|
| 0 | 359.5 | 295.0 | 132.4 | 76.3 | 68.0 |
| 0.25 | 353.1 | 302.7 | 139.3 | 75.0 | 74.0 |
| 0.5 | 350.7 | 307.5 | 143.2 | 79.1 | 76.3 |
| 0.75 | 344.6 | 304.4 | 138.8 | 86.6 | 81.0 |
| 1 | 332.6 | 287.4 | 121.3 | 86.7 | 73.7 |

## Ti$_3$(Sn,Al)C$_2$

### Elastic constants

Calculated enthalpy-of-formation and mechanical stability criteria suggested[159] that all of the Ti$_3$Sn$_x$Al$_{1-x}$C$_2$ solid solutions (table 50), where x ranged from 0 to 1, were thermodynamically and elastically stable. The elastic constants (table 51) and elastic moduli (table 52) indicated that all of the compounds were brittle and isotropic, with relatively high melting points (table 53).

*Table 52. Elastic moduli (GPa) of Ti$_3$(Sn$_x$Al$_{1-x}$)C$_2$*

| x | B | G |
|------|-------|-------|
| 0 | 159.1 | 134.4 |
| 0.25 | 135.7 | 135.4 |
| 0.5 | 163.1 | 135.7 |
| 0.75 | 165.0 | 131.0 |
| 1 | 157.3 | 120.9 |

*Table 53. Densities (g/cm$^3$) and melting-points (K) of Ti$_3$(Sn$_x$Al$_{1-x}$)C$_2$*

| x | Density | Melting-Point |
|------|--------|--------|
| 0 | 4.222 | 1875.0 |
| 0.25 | 4.667 | 1867.4 |
| 0.5 | 5.100 | 1867.4 |
| 0.75 | 5.517 | 1844.4 |
| 1 | 5.917 | 1782.9 |

### Ti$_2$SnC

#### *Elastic constants*

First-principles calculations were used[160] to study the structural (figure 36) and elastic (table 54) properties of M$_2$SnC phases, where M was titanium, zirconium, niobium or hafnium. The effect of pressures of up to 20GPa upon the lattice constants was to cause contractions along the a-axis which were greater than those along the c-axis. There was a quadratic dependence of the lattice parameters upon the applied pressure. The elastic constants and their pressure dependences (table 55) were calculated by using the static finite-strain technique, and a linear dependence of the elastic stiffness upon pressure was found. Bulk, shear and Young's moduli, together with the Poisson ratio were derived (table 56) for ideal polycrystalline M$_2$SnC aggregates.

*Figure 36. Structure of Ti$_2$SnC*
*(a = 3.16Å, c = 13.68Å, theoretical density = 6.36g/cm$^3$)*
*grey: carbon, orange: titanium, green: tin*

*Table 54. Calculated elastic constants of $M_2SnC$-type phases*

| Phase | $C_{11}$ (GPa) | $C_{33}$ (GPa) | $C_{44}$ (GPa) | $C_{12}$ (GPa) | $C_{13}$ (GPa) | $C_{66}$ (GPa) |
|---|---|---|---|---|---|---|
| $Ti_2SnC$ | 303 | 308 | 121 | 84 | 88 | 109 |
| $Zr_2SnC$ | 279 | 272 | 111 | 70 | 89 | 104 |
| $Nb_2SnC$ | 315 | 309 | 124 | 99 | 141 | 108 |
| $Hf_2SnC$ | 311 | 306 | 119 | 92 | 97 | 109 |

*Table 55. Pressure dependences of the elastic constants of $M_2SnC$-type phases*

| Phase | $\partial B/\partial P$ | $\partial C_{11}/\partial P$ | $\partial C_{33}/\partial P$ | $\partial C_{44}/\partial P$ | $\partial C_{12}/\partial P$ | $\partial C_{13}/\partial P$ |
|---|---|---|---|---|---|---|
| $Ti_2SnC$ | 4.26 | 6.78 | 5.48 | 2.72 | 2.16 | 3.62 |
| $Zr_2SnC$ | 4.1 | 4.54 | 6.26 | 2.24 | 3.5 | 3.6 |
| $Nb_2SnC$ | 4.48 | 4.78 | 6.52 | 2.9 | 3.44 | 4.64 |
| $Hf_2SnC$ | 4.26 | 4.66 | 6.0 | 2.78 | 3.84 | 3.88 |

*Table 56. Calculated elastic moduli of $M_2SnC$-type phases*

| Phase | B (GPa) | G (GPa) | E (GPa) | Poisson Ratio |
|---|---|---|---|---|
| $Ti_2SnC$ | 159.5 | 113.6 | 275.4 | 0.212 |
| $Zr_2SnC$ | 147.3 | 103.8 | 252.1 | 0.215 |
| $Nb_2SnC$ | 188.7 | 106.8 | 188.7 | 0.262 |
| $Hf_2SnC$ | 166.7 | 12.1 | 274.8 | 0.225 |

*Table 57. Elastic properties (GPa) of $M_3SnC_2$ phases*

| Phase | $C_{11}$ | $C_{33}$ | $C_{44}$ | $C_{66}$ | $C_{12}$ | $C_{13}$ | Poisson Ratio |
|---|---|---|---|---|---|---|---|
| $Ti_3SnC_2$ | 321 | 304 | 115 | 110 | 100 | 79 | 0.215 |
| $Zr_3SnC_2$ | 280 | 257 | 110 | 94 | 92 | 84 | 0.227 |
| $Hf_3SnC_2$ | 320 | 300 | 115 | 113 | 95 | 96 | 0.227 |

**Ti₃SnC₂**

### Hardness

Density functional theory calculations were used[161] to determine the structural and elastic properties (table 57, figure 37) of tin-containing 312-type phases of the form, $M_3SnC_2$, where M was titanium, zirconium or hafnium. The *a* lattice constant increased as M was changed from titanium to hafnium. The **a** lattice constant of $Ti_3SnC_2$ is 3.14Å, the **c** lattice constant is 18.65Å and the theoretical density is 5.95g/cm³. The hafnium-based material was elastically almost isotropic. The covalency of the M-C bonds increased as M was switched from titanium to hafnium, via zirconium. The order of machinability was predicted to be: $Zr_3SnC_2 > Hf_3SnC_2 > Ti_3SnC_2$.

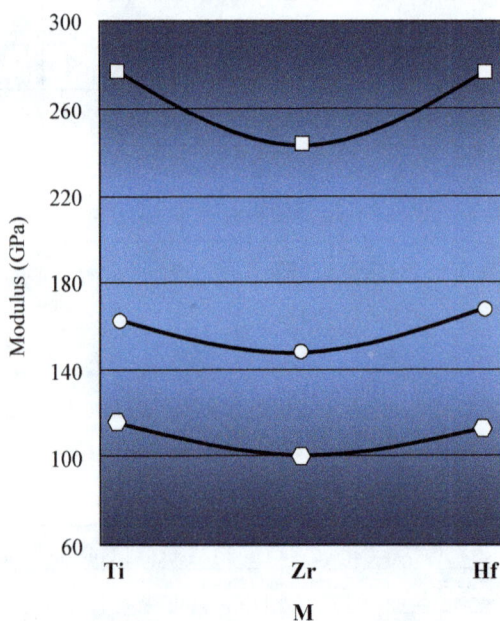

*Figure 37. Elastic moduli of M₃SnC₂ phases. Squares: Young's modulus, circles: bulk modulus, hexagons: shear modulus*

## $(Ti,V)_2AlC$

### *Hardness*

The effect of vanadium upon the properties of $(Ti_{1-x}V_x)_2AlC$ solid solutions, where x was 0, 0.25, 0.50, 0.75 or 1.00, was examined[162] under up to 50GPa compression by means of density functional theory. The structures belonged to the hexagonal $P6_3/mmc$ space group, with 8 atoms in a unit cell which contained 2 molecules. The **a** and **c** lattice constants decreased with increasing vanadium content due to the ionic-size difference between titanium (0.67Å) and vanadium (0.64Å). The value of **a** decreased more markedly than that of **c** and this reflected the fact that **a** is more easily adjusted than is **c** when the titanium content is increased from 0 to 0.75. The optimized lattice constants agreed well with experimental values, and decreased with increasing vanadium content (table 58). The solid solutions were compressible along both the *a* and *c* directions at pressures of 0 to 50GPa. The hardness attained an appreciable maximum value at a vanadium content of 75% (table 59). The $C_{44}$ constant, which was directly related to the valence-electron concentration, increased with increasing vanadium content (table 60). The bulk, shear and Young's moduli, attained an appreciable maximum value at a vanadium content of 75% (table 61). Shear-anisotropy factors showed that the compounds which contained titanium and vanadium were more anisotropic than was $Ti_2AlC$, and tended to have an improved ductility.

*Table 58. Lattice parameters (Å) of $(Ti_{1-x}V_x)_2AlC$*

| x | a | c | c/a |
|---|---|---|---|
| 0.00 | 3.062 | 13.643 | 4.456 |
| 0.25 | 3.013 | 13.593 | 4.511 |
| 0.50 | 2.973 | 13.473 | 4.532 |
| 0.75 | 2.937 | 13.322 | 4.536 |
| 1.00 | 2.905 | 13.140 | 4.523 |

### *Elastic constants*

The bulk modulus, shear modulus, Young's modulus and Debye temperature increased with increasing pressure up to 50GPa. Analysis of the elastic constants showed that, when the vanadium content attained 75%, the stiffness was greatly increased as compared with that of Ti$_2$AlC. The energy band structure and density-of-states indicated that the materials possessed the characteristics of a metal; due mainly to contributions arising from Ti-d, Ti-d, V-d and V-d orbitals. When the vanadium content was 75%, the bulk modulus, shear modulus, Young's modulus and Vickers hardness were 11.8, 8.4, 9.1 and 85.1%, respectively, higher than the values for Ti$_2$AlC compound.

*Table 59. Microhardness parameter and Poisson ratio of (Ti$_{1-x}$V$_x$)$_2$AlC*

| x | Hardness Parameter | Poisson Ratio |
|---|---|---|
| 0.00 | 26.45 | 1.07 |
| 0.25 | 23.65 | 1.17 |
| 0.50 | 24.32 | 1.33 |
| 0.75 | 27.93 | 1.20 |
| 1.00 | 22.59 | 1.42 |

*Table 60. Elastic constants (GPa) of (Ti$_{1-x}$V$_x$)$_2$AlC*

| x | C$_{11}$ | C$_{12}$ | C$_{13}$ | C$_{33}$ | C$_{44}$ |
|---|---|---|---|---|---|
| 0.00 | 306 | 59 | 66 | 283 | 122 |
| 0.25 | 294 | 54 | 84 | 282 | 126 |
| 0.50 | 300 | 73 | 86 | 274 | 134 |
| 0.75 | 331 | 70 | 85 | 308 | 141 |
| 1.00 | 333 | 44 | 112 | 339 | 159 |

*Table 61. Elastic moduli (GPa) of (Ti$_{1-x}$V$_x$)$_2$AlC*

| x | B | G | E |
|---|---|---|---|
| 0.00 | 142 | 120 | 281 |
| 0.25 | 146 | 115 | 272 |
| 0.50 | 151 | 118 | 280 |
| 0.75 | 161 | 131 | 309 |
| 1.00 | 152 | 112 | 283 |

## V₂AlC

### Hardness

The elastic constants and hardness of V$_2$AlC (a = 2.91Å, c = 13.10Å, theoretical density = 4.87g/cm$^3$) and Cr$_2$AlC (a = 2.86, c = 12.8, theoretical density = 5.24g/cm$^3$) single crystals were determined[163] by micro-/nano-indentation. The C$_{33}$ and C$_{11}$ values were determined by using a Berkovich tip and were all found to lie between 320 and 350GPa. The results confirmed that many MAX phases are elastically relatively isotropic. Hardness values which were obtained by using Vickers, Berkovich or a 5μm spherical tip, on the two orthogonal Cr$_2$AlC surfaces, were comparable; with an average of 9.0GPa. The hardness was at most 20% higher when the basal planes were loaded along [00•1] than when loaded edge-on. The Cr$_2$AlC surfaces exhibited an average micro yield-stress of 2.7GPa, while less-defective V$_2$AlC crystals sustained stresses which were of the order of 20GPa. In some 60% of cases, there were pop-ins; some of which were substantial. Post-indentation scanning electron microscopy clearly revealed the plastic anisotropy of the crystals. Large pile-ups which were located near to the indent edges and delamination cracks, after loading along the [00•1] and [10•0] directions respectively, were consistent with deformation via ripplocation.

*Table 62. Strength of V₂GaC as a function of grain size*

| Grain Diameter (μm) | Grain Thickness (μm) | Test | Strength (MPa) |
|---|---|---|---|
| 49 | 19 | 3-point flexure | 263 |
| 49 | 19 | compression | 740 |
| 108 | 37 | 3-point flexure | 290 |
| 108 | 37 | compression | 600 |
| 119 | 47 | 3-point flexure | 270 |
| 119 | 47 | compression | 530 |
| 405 | 106 | 3-point flexure | 61 |
| 405 | 106 | compression | 390 |

Samples were prepared[164] from vanadium and $V_2O_5$ by using spark plasma sintering. The vanadium and $V_2O_5$, plus a carbon source and aluminium, were subjected to high-energy mixing for 600s before sintering at 1350C under pressures of 10 or 30MPa. The predominant phase was $V_2AlC$ in samples which were synthesized using vanadium powder while, in the case of samples prepared using $V_2O_5$, there was some formation of $Al_2O_3$. Lath-like and spherical $V_2AlC$ phases were found in samples which contained vanadium and $V_2O_5$, respectively. A bending strength of 347MPa was found for samples which consisted of $V_2AlC$. When samples were prepared[165] by microwave sintering of elemental powder mixtures at 1200, 1300 or 1400C, the highest bending strength (189MPa) was found for specimens which were sintered at 1300C, while the highest Vickers hardness (892Hv) was found for samples which were sintered at 1400C.

Low-energy ion bombardment was used to prepare[166] high-density $V_2AlC$ coatings by magnetron sputtering, and the mechanical properties were determined by nano-indentation. Increasing the bombarding-ion energy from 15 to 35eV densified the coating and reduced the surface roughness from 20.2 to 11.9nm, while increasing the hardness from 14 to 21GPa. The Young's modulus was increased from 309 to 363GPa. When the bombarding-ion energy was increased to 50eV, the aluminium content was markedly decreased and the coating changed from $V_2AlC$ to a $V_2C$ plus VC mixture.

### *Elastic constants*

First-principles calculations were made[167] of the elastic constants (table 62) and Vickers hardness of $V_2AlC$ (3.65GPa) and $V_2GaC$ (3.10GPa). The temperature- and pressure-dependences of the bulk modulus were also estimated, using the quasi-harmonic Debye model, between 0 and 1000K (figure 38) and between 0 and 50GPa (figure 39). The bulk modulus was greater for the $V_2AlC$, and directional bonding between the atoms of these compounds was deduced to be strong due to their large shear moduli. The relatively large Young's moduli (table 63) of these materials suggested that they should be stiff. The shear/bulk ratios of the moduli were 0.79 and 0.68 for $V_2AlC$ and $V_2GaC$, respectively; indicating that both of them were brittle. This was also suggested by the empirical rule that a Poisson ratio of 0.26 demarcates brittle and ductile behaviours. The Poisson ratios were 0.18 and 0.22 for $V_2AlC$ and $V_2GaC$, respectively; again suggesting brittleness. An elastic anisotropy of the crystals was suggested to explain a tendency to exhibit microcracking. The anisotropy values of $V_2AlC$ and $V_2GaC$ were 1.04 and 1.09, respectively. The compressibility along the c-axis was smaller than that along the a-axis in the case of $V_2AlC$, while the reverse was true of $V_2GaC$. Calculations indicated that the V-C bonds were more strongly covalent than were the V-Al or V-Ga bonds in these materials.

94

*Table 63. Elastic constants of V₂GaC*

| C$_{11}$ (GPa) | C$_{12}$ (GPa) | C$_{13}$ (GPa) | C$_{33}$ (GPa) | C$_{44}$ (GPa) | Reference |
|---|---|---|---|---|---|
| 346 | 71 | 106 | 314 | 151 | [168] |
| 338 | 92 | 148 | 328 | 155 | see under Cr₂AlC |
| 339 | 71 | 100 | 319 | 148 | [169] |

*Table 64. Elastic moduli of V₂Al and V₂GaC*

| Phase | B (GPa) | G (GPa) | E (GPa) | Poisson Ratio |
|---|---|---|---|---|
| V₂AlC | 169.7 | 135.7 | 321.0 | 0.18 |
| V₂GaC | 182.6 | 124.4 | 307.0 | 0.22 |

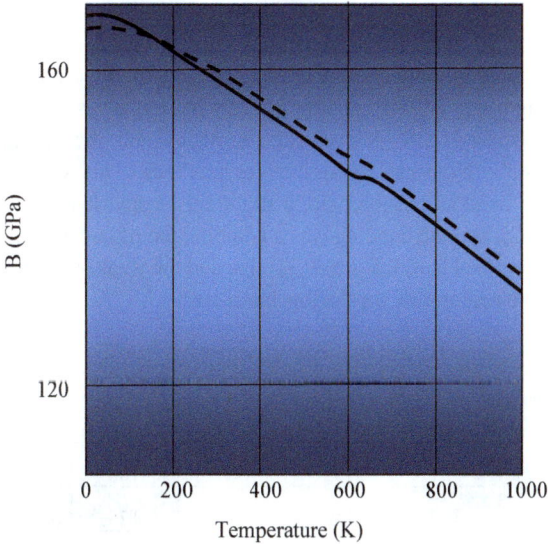

*Figure 38. Temperature dependences (0GPa) of the bulk moduli of V₂AlC (broken line) and V₂GaC (solid line)*

The strength decreased, as usual, with increasing grain size[170] (table 64). All of the $V_2AlC$ samples failed via the formation of shear bands. In general, when polycrystalline MAX phases are loaded in compression they either fail in a brittle way or via the formation of shear bands that are inclined at between 25 and 40° to the loading direction. The former mode is more common when the strain-rate is high and the grain-size is small. In view of the layered nature of MAX phases it is not surprising that a preferred failure mode is shear. The Hall–Petch relationship is obeyed regardless of the failure mode. The tendency of MAX phases to fail in compression via the formation of shear bands suggests that they have a low shear resistance, and experience shows that punch shear tests provide the best estimate of shear resistance.

## $V_4AlC_3$

### *Hardness*

The effect of spark plasma sintering upon the microstructure and mechanical properties was determined[171] for samples which were made from vanadium, aluminium and carbon-black powders, and where VC and $Al_2O_3$ were side-products. The $V_4AlC_3$ was lath-like, and the fracture surfaces of bend-test specimens exhibited deformed and separate lath-like phases. Deformed areas revealed various crack-propagation mechanisms such as separation and branching, deflection and bridging, plus particle pull-out and breakage. The bending strength was 389MPa and the Vickers hardness was 6.74GPa for material of 99% theoretical density. The mechanical properties and theoretical hardness were originally predicted[172] by using first-principles methods. This showed that it resisted shape changes and uniaxial tension, with a slight anisotropy. It was more brittle than α-$Nb_4AlC_3$ or $Ta_4AlC_3$, and the bonding was a combination of covalent, ionic and metallic. The calculated theoretical hardness was 9.33GPa, and the weaker covalent bonding of Al-V was deemed to be responsible for the low hardness.

*Table 65. Elastic constants of $V_4AlC_3$*

| $C_{11}$ (GPa) | $C_{12}$ (GPa) | $C_{13}$ (GPa) | $C_{33}$ (GPa) | $C_{44}$ (GPa) | Reference |
|---|---|---|---|---|---|
| 435 | 121 | 105 | 384 | 168 | [173] |
| 458 | 107 | 110 | 396 | 175 | present work |

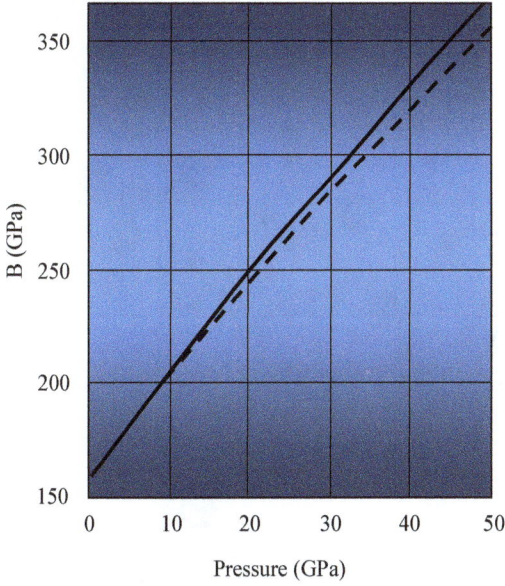

*Figure 39. Pressure dependences (300K) of the bulk moduli of $V_2AlC$ (broken line) and $V_2GaC$ (solid line)*

### Elastic constants

The constants which were derived in the present work differed slightly from those found elsewhere (table 65).

*Table 65. Elastic constants of $V_4AlC_3$*

| $C_{11}$ (GPa) | $C_{12}$ (GPa) | $C_{13}$ (GPa) | $C_{33}$ (GPa) | $C_{44}$ (GPa) | Reference |
|---|---|---|---|---|---|
| 435 | 121 | 105 | 384 | 168 | 173 |
| 458 | 107 | 110 | 396 | 175 | present work |

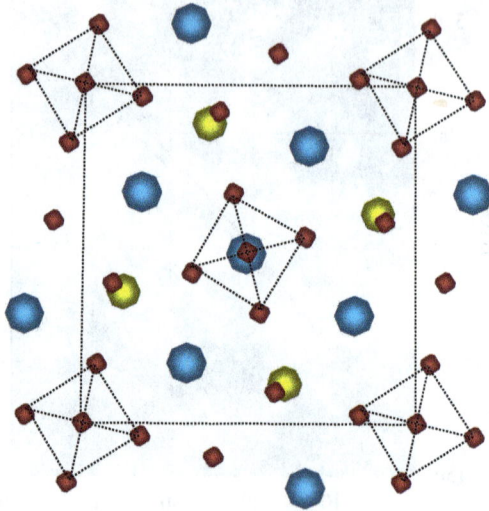

*Figure 40. Projection of atoms on the (001) plane of $Y_5Si_2B_8$*
*brown: boron, blue: yttrium, yellow: silicon*

*Table 66. Comparison of the elastic constants (GPa) of the*
*$Y_5Si_2B_8$ MAB phase with those of various MAX phases*

| Phase | $C_{11}$ | $C_{33}$ | $C_{44}$ | $C_{66}$ | $C_{12}$ | $C_{13}$ |
|-------|------|------|------|------|------|------|
| $Y_5Si_2B_8$ | 312 | 245 | 100 | 106 | 71 | 53 |
| $Ti_2AlC$ | 304 | 274 | 119 | 127 | 50 | 57 |
| $Ti_3AlC_2$ | 355 | 292 | 123 | 142 | 71 | 68 |
| $Ti_3SiC_2$ | 354 | 344 | 165 | 131 | 91 | 103 |

## $Y_5Si_2B_8$

### Elastic constants

This MAB phase (figure 40), which consisted of alternating $YB_4$ and $Y_3Si_2$ slabs in the [001] direction, was investigated[174] in comparison with MAX phases (tables 66 and 67). Density functional theory calculations revealed that the MAB phase had bonding and elastic properties which were similar to those of MAX phases. Strong covalent bonds in the two-dimensional boron network on the (001) plane in the $YB_4$ slabs, and between silicon atoms on the (002) plane in the $Y_3Si_2$ slabs, explained a high stiffness (288GPa) of the $Y_5Si_2B_8$ in the ab-plane. Weak Y2-Si and Y1-B2 bonds which connected the $YB_4$ and $Y_3Si_2$ slabs explained a low (200GPa) Young's modulus in the [001] direction. A low shear-deformation resistance was attributed to the existence of metallic bonding, and to weak bonding within the $B_6$ octahedra. The possible slip systems were {001}<100> and {110}<111>. The Pugh ratio and low (104GPa) shear modulus implied that this MAB phase would be damage-tolerant.

*Table 67. Comparison of the moduli of the $Y_5Si_2B_8$*
*MAB phase with those of various MAX phases*

| Phase | B (GPa) | G (GPa) | E (GPa) | Poisson Ratio |
|---|---|---|---|---|
| $Y_5Si_2B_8$ | 130 | 104 | 246 | 0.184 |
| $Ti_2AlC$ | 135 | 120 | 278 | 0.157 |
| $Ti_3AlC_2$ | 157 | 130 | 306 | 0.176 |
| $Ti_3SiC_2$ | 183 | 141 | 337 | 0.193 |

## $Zr_2AlC$

### Hardness

When a stoichiometric mixture of high-purity $ZrH_2$ and carbon powders in the molar ratio of 2:1 was subjected to intense milling, and infiltrated with molten aluminium for 5h, the resultant phase[175] had a Vickers hardness of 4.5GPa and fracture toughness of $23MPam^{0.5}$. First-principles density functional theory calculations of the Vickers hardness have suggested that the bulk modulus increases and the shear modulus decreases upon partial bismuth, tin or antimony replacement of the aluminium.

*Table 68. Calculated elastic constants of Zr-Al-C phases*

| Phase | $C_{11}$ (GPa) | $C_{12}$ (GPa) | $C_{13}$ (GPa) | $C_{33}$ (GPa) | $C_{44}$ (GPa) | $C_{66}$ (GPa) |
|---|---|---|---|---|---|---|
| $Zr_2AlC$ | 265.14 | 68.29 | 66.51 | 227.00 | 98.43 | 97.97 |
| $Zr_3AlC_2$ | 313.96 | 79.03 | 75.35 | 252.18 | 117.47 | 102.30 |
| $Zr_2Al_3C_4$ | 402.97 | 115.48 | 93.36 | 345.31 | 143.74 | 165.41 |
| $Zr_3Al_3C_5$ | 411.68 | 117.52 | 97.69 | 359.64 | 147.08 | 167.60 |
| $Zr_4Al_3C_6$ | 420.09 | 115.86 | 101.39 | 370.06 | 152.11 | 171.69 |
| $Zr_2Al_4C_5$ | 368.96 | 121.78 | 84.24 | 364.91 | 123.59 | 128.52 |
| $Zr_3Al_4C_6$ | 379.95 | 124.70 | 90.19 | 377.16 | 127.63 | 131.13 |

*Table 69. Calculated moduli of Zr-Al-C phases*

| Phase | B (GPa) | G (GPa) | E (GPa) | Poisson Ratio |
|---|---|---|---|---|
| $Zr_2AlC$ | 128.54 | 95.77 | 230.14 | 0.20 |
| $Zr_3AlC_2$ | 148.04 | 107.36 | 259.38 | 0.20 |
| $Zr_2Al_3C_4$ | 194.23 | 150.95 | 359.68 | 0.21 |
| $Zr_3Al_3C_5$ | 200.31 | 153.98 | 367.72 | 0.19 |
| $Zr_4Al_3C_6$ | 204.75 | 158.07 | 377.15 | 0.19 |
| $Zr_2Al_4C_5$ | 186.82 | 130.00 | 316.57 | 0.22 |
| $Zr_3Al_4C_6$ | 193.96 | 133.18 | 325.13 | 0.22 |

The calculated elastic moduli showed that the modified materials were more anisotropic than $Zr_2AlC$ and tended to have an increased ductility, while their Vickers hardness was decreased[176]. Theoretical Vickers hardness values were also predicted[177] for $Zr_2AlC$, $Zr_2SiC$, $Zr_2PC$ and $Zr_2SC$ by using first-principles calculations which were based upon density functional theory and the plane-wave pseudopotential method. The theoretical hardness was estimated by calculating Mulliken bond populations and the electronic density of states. Thermodynamic properties such as the temperature and pressure dependence of the bulk modulus were deduced by using the quasi-harmonic Debye model.

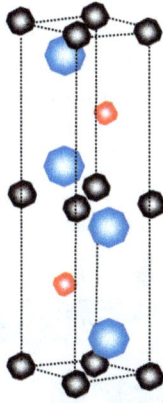

*Figure 43. Crystal structure of Zr₂AlC*
*grey: carbon, blue: zirconium, red: aluminium*

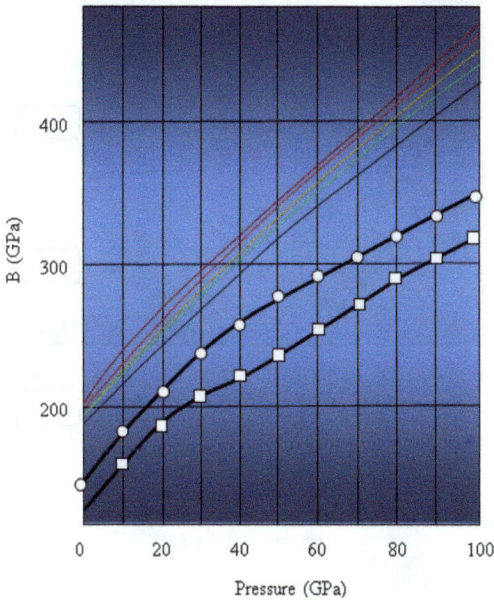

*Figure 41. Bulk modulus of Zr-Al-C phases as a function of pressure. Brown line:*
*Zr₄Al₃C₆, red line: Zr₃Al₃C₅, yellow line: Zr₂Al₃C₄, green line: Zr₃Al₄C₆, black line:*
*Zr₂Al₄C₅, circles: Zr₃Al₁C₂, squares: Zr₂AlC*

### Elastic constants

First-principles calculations were used[178] to determine the elastic properties (table 68) of Zr–Al–C nanolaminates. The nanolaminates exhibited crystallographic similarities, but their properties were very different. The predicted moduli (table 69) increased in the order: $Zr_2AlC < Zr_3AlC_2 < Zr_2Al_4C_5 < Zr_3Al_4C_6 < Zr_2Al_3C_4 < Zr_3Al_3C_5 < Zr_4Al_3C_6$. In the presence of an external pressure, the bulk (figure 41) and shear (figure 42) moduli exhibited a linear response, except in the case of $Zr_2AlC$ (figure 43) and $Zr_3AlC_2$ (figure 44). These were MAX phases.

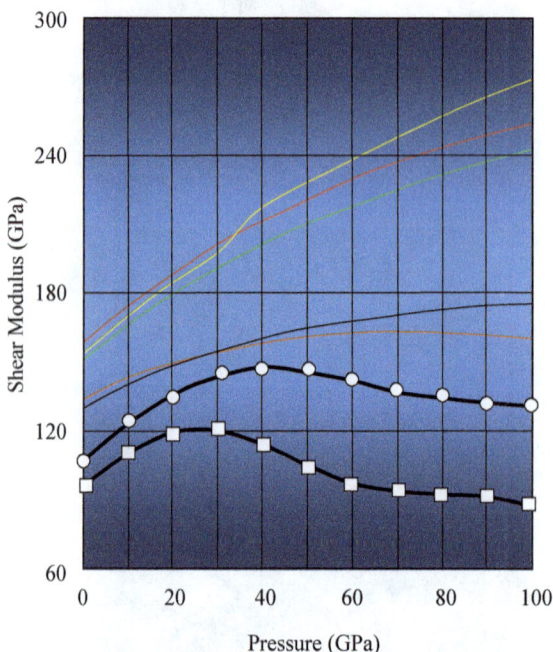

*Figure 42. Shear modulus of Zr-Al-C phases as a function of pressure. Yellow line: $Zr_3Al_3C_5$, red line: $Zr_4Al_3C_6$, black line: $Zr_2Al_3C_4$, orange line: $Zr_3Al_4C_6$, circles: $Zr_3AlC_2$, squares: $Zr_2AlC$*

The effect of vanadium substitution upon $Zr_2AlC$ was predicted[179] by using first-principles methods. The resultant lattice parameters (table 70), elastic constants (table 71), elastic moduli (table 72) and marked anisotropy factors indicated that the compressibility along the a-axis was greater than that along the c-axis. The latter changes gradually decreased as the vanadium concentration increased. The macroscopic mechanical properties implied that the substituted materials were brittle, stiff and hard and exhibited metallic behaviour. The latter was attributed to strong p-d covalent bonding. The densities and melting-points varied smoothly between the end-members (table 73).

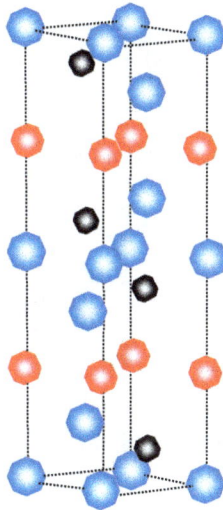

*Figure 44. Crystal structure of $Zr_3AlC_2$*
*blue: carbon, red: zirconium, grey: aluminium*

*Table 70. Lattice parameters (Å) and bulk modulus (GPa) of vanadium-alloyed $Zr_2AlC$*

| Compound | a-axis | c-axis | B |
|---|---|---|---|
| $Zr_2AlC$ | 3.27 | 14.73 | 161.73 |
| $(Zr_{0.75}V_{0.25})_2AlC$ | 3.19 | 14.12 | 168.44 |
| $(Zr_{0.50}V_{0.50})_2AlC$ | 3.09 | 13.80 | 177.03 |
| $(Zr_{0.25}V_{0.75})_2AlC$ | 2.98 | 13.37 | 196.40 |
| $V_2AlC$ | 2.85 | 12.96 | 230.26 |

Materials Research Forum LLC
https://doi.org/10.21741/9781644901274

*Table 71. Elastic constants (GPa) of vanadium-alloyed $Zr_2AlC$*

| Compound | $C_{11}$ | $C_{12}$ | $C_{13}$ | $C_{33}$ | $C_{55}$ | $C_{66}$ |
|----------|----------|----------|----------|----------|----------|----------|
| $Zr_2AlC$ | 261.91 | 63.55 | 84.40 | 260.56 | 103.27 | 99.18 |
| $(Zr_{0.75}V_{0.25})_2AlC$ | 301.05 | 79.39 | 115.83 | 295.81 | 137.64 | 110.83 |
| $(Zr_{0.50}V_{0.50})_2AlC$ | 324.91 | 84.27 | 115.72 | 332.11 | 158.49 | 120.32 |
| $(Zr_{0.25}V_{0.75})_2AlC$ | 346.80 | 105.72 | 150.60 | 343.14 | 163.18 | 120.53 |
| $V_2AlC$ | 389.78 | 126.64 | 140.06 | 377.74 | 181.40 | 131.57 |

*Table 72. Elastic moduli (GPa) and Vickers hardness (GPa) of vanadium-alloyed $Zr_2AlC$*

| Compound | G | E | $H_v$ | Poisson Ratio |
|----------|---|---|-------|---------------|
| $Zr_2AlC$ | 97.67 | 237.31 | 16.356 | 0.21 |
| $(Zr_{0.75}V_{0.25})_2AlC$ | 114.48 | 280.08 | 17.344 | 0.22 |
| $(Zr_{0.50}V_{0.50})_2AlC$ | 129.87 | 313.74 | 20.685 | 0.20 |
| $(Zr_{0.25}V_{0.75})_2AlC$ | 128.07 | 318.06 | 16.687 | 0.24 |
| $V_2AlC$ | 146.53 | 359.53 | 20.120 | 0.23 |

*Table 73. Densities and melting-points of vanadium-alloyed $Zr_2AlC$*

| Compound | Density (g/cm$^3$) | Melting Point (K) |
|----------|--------------------|-------------------|
| $Zr_2AlC$ | 5.3619 | 1531 |
| $(Zr_{0.75}V_{0.25})_2AlC$ | 5.3119 | 1701 |
| $(Zr_{0.50}V_{0.50})_2AlC$ | 5.2596 | 1827 |
| $(Zr_{0.25}V_{0.75})_2AlC$ | 5.2068 | 1909 |
| $V_2AlC$ | 5.1339 | 2090 |

## Zr₃AlC₂

### Hardness

A study[180] of the synthesis and structure of this phase, as prepared by the reactive hot-pressing of $ZrH_2$, aluminium and carbon powders, showed that the crystal structure was hexagonal; with the space group, $P6_3/mmc$. The a- and c- lattice parameters were 3.33308 and 19.9507Å, respectively. The Vickers hardness, using a force of 30N, was 4.4GPa.

## Zr₂Al₃C₅

### Elastic Constants

The equations of state were determined[181] by using the first-principles pseudopotential total energy method. The chemical bonding and layer characteristics were similar to those of nanolaminates such as $Ti_2AlC$ and $Ti_3AlC_2$, and the present material exhibited strong covalent bonding among interleaved Al-C-Zr-C-Zr-C-Al chains. It was expected to be easily machinable and damage-tolerant, and the theoretical bulk modulus was 160GPa. Elastic anisotropy existed under pressures of less than 20GPa. The bulk modulus of this phase was about 70% of that of ZrC; estimated to be 229GPa. The pressure derivative of the bulk modulus was 5.2. The axial c/a ratio decreases rapidly as the pressure is increased from 0 to 20GPa, and saturates at a constant value of 7.52. The trend in the c/a ratio shows that the c-parameter contracts more rapidly, than does the a-parameter, below 20GPa. The elastic stiffness along the c-axis is lower than that along the basal plane when $Zr_2Al_3C_5$ is under pressure. The elastic anisotropy indicates that the intraplanar atomic bonds are stronger than the interplane bonds. A similar trend is observed in $Ti_3SiC_2$. In $Zr_2Al_3C_5$, the $AlC_2$ block is very weakly bonded to the $Zr_2Al_2C_3$ units. Due to the weak interplanar bonding, the material can undergo deformation via basal plane slip, cleavage and delamination; leading to microscale plasticity, easy machinability and damage tolerance.

## Zr₂(Al,Bi)C

### Hardness

The structural, electronic and optical properties of $Zr_2(Al_{1-x}Bi_x)C$ showed[182] that the inclusion of bismuth on the aluminium site led to an increase in the $a$ lattice constant, while the $c$ lattice constant decreased upon increasing x up to 0.58. The $c$ values were more affected than were the $a$ values when aluminium was replaced by bismuth; implying that the $c$-value was more dependent upon the M-A bonds than upon other bonds. The calculated band structures also implied that the electrical conductivity along the $c$ direction should be small when compared with that in the *ab*-plane. The total density-of-states at the Fermi level increased almost linearly with increasing bismuth contents, x, of between 0.25 and 0.75. The Mulliken atomic populations indicated that the Zr–C bonds were more covalent in $Zr_2BiC$ than in $Zr_2AlC$. The calculated Vickers hardness values of $Zr_2AlC$ and $Zr_2BiC$ were 5.96 and 1.94GPa, respectively; suggesting that $Zr_2BiC$ was relatively soft and easier to machine.

## (Zr,Nb)₂AlC

### Hardness

Density functional theory was used[183] to investigate the properties of $Zr_2AlC$ and $Nb_2AlC$, and especially those of their solid solutions, $(Zr_{1-x}Nb_x)_2AlC$. The partial inclusion of niobium on the M-site improved the bond strength and the hardness of $Zr_2AlC$. The stiffness of the solid solutions increased with x, and improved its ability to resist shear deformation. Given their large negative Cauchy pressures, all of the compositions were predicted to exhibit directional covalent bonding. The composition with x = 0.2 was expected to be more brittle than the others, while the Debye temperature was predicted to be highest for the composition with x = 0.4. Covalency increased with increasing x-value, and was suggested to explain an increased stiffness.

## (Zr,Ti)₃AlC₂

### Hardness

Samples of $(Zr_{1-x}Ti_x)_3AlC_2$ were prepared[184], and their mechanical properties were investigated using density functional theory. The predicted lattice parameters were in good agreement with experimentally determined ones, and deviate by less than 0.5% from Vegard's law. The $Ti_3AlC_2$ end-member had a higher predicted Vickers hardness than that of $Zr_3AlC_2$, in agreement with experimental data.

*Table 74. Calculated elastic constants of $(Zr_{1-x}Ti_x)_3AlC_2$*

| x | $C_{11}$ (GPa) | $C_{12}$ (GPa) | $C_{13}$ (GPa) | $C_{33}$ (GPa) | $C_{44}$ (GPa) | $C_{66}$ (GPa) |
|---|---|---|---|---|---|---|
| 0 | 308.59 | 89.33 | 97.37 | 318.24 | 82.27 | 109.63 |
| 0.5 | 313.41 | 90.86 | 97.20 | 331.38 | 89.09 | 111.28 |
| 1 | 358.86 | 99.95 | 92.33 | 366.22 | 102.19 | 129.45 |

### Elastic constants

The structural and mechanical properties of $(Zr_{1-x}Ti_x)_3AlC_2$ phases were investigated[185] by using the full-potential plane-wave method. These materials exhibited a metallic behavior which was attributed to p-d hybridization, and they were mechanically stable. The results showed that $C_{33}$ was larger than $C_{11}$ for every sample; indicating that the a- and b-axes were more compressible than was the c-axis. The results could be explained in terms of the existence of strong covalent bonding in the [001] direction. It was noted that the $C_{11}$ and $C_{33}$ constants were much higher than the others, reflecting an elastic anisotropy of the phases. As x increased, the bulk, shear and Young's moduli increased (tables 74 and 75). The $Ti_3AlC_2$ exhibited a greater ability to resist deformation. The opposite was true in the case of $Zr_3AlC_2$. The calculated values of the Poisson ratio were 0.255, 0.249 and 0.233 for $Zr_3AlC_2$, $(Zr_{0.5}Ti_{0.5})_3AlC_2$ and $Ti_3AlC_2$, respectively. This suggested that the chemical bonding was more ionic for $Zr_3AlC_2$, have mixed for $(Zr_{0.5}Ti_{0.5})_3AlC_2$ and more covalent for $Ti_3AlC_2$. All of the Poisson ratios were less than 0.26, and were therefore all brittle in nature. The B/G ratios were less than 1.75; again confirming that the phases were brittle. The hardness was lowest for $Zr_3AlC_2$ and greatest for $Ti_3AlC_2$. The elastic constants increased when the pressure was increased. The bulk, shear and Young's moduli increased almost linearly with pressure increases.

*Table 75. Moduli of $(Zr_{1-x}Ti_x)_3AlC_2$*

| x | B (GPa) | G (GPa) | E (GPa) | Poisson Ratio |
|---|---|---|---|---|
| 0 | 167.01 | 97.33 | 244.50 | 0.255 |
| 0.5 | 169.76 | 102.11 | 255.16 | 0.249 |
| 1 | 183.68 | 119.10 | 293.80 | 0.233 |

# About the Author

**Dr. Fisher** has wide knowledge and experience of the fields of engineering, metallurgy and solid-state physics, beginning with work at Rolls-Royce Aero Engines on turbine-blade research, related to the Concord supersonic passenger-aircraft project, which led to a BSc degree (1971) from the University of Wales. This was followed by theoretical and experimental work on the directional solidification of eutectic alloys having the ultimate aim of developing composite turbine blades. This work led to a doctoral degree (1978) from the Swiss Federal Institute of Technology (Lausanne). He then acted for many years as an editor of various academic journals, in particular *Defect and Diffusion Forum*. In recent years he has specialized in writing monographs which introduce readers to the most rapidly developing ideas in the fields of engineering, metallurgy and solid-state physics. He is co-author of the widely-cited student textbook, *Fundamentals of Solidification*. Google Scholar credits him with 7522 citations and a lifetime h-index of 12.

# Materials Index

$V_2GeC$, 14, 16, 74
$V_2PC$, 39
$V_2SnC$, 38
$V_4AlC_3$, 96, 97

$W_2GeC$, 14, 16

$Y_5Si_2B_8$, 98, 99

$Zr_2Al_2C_3$, 105
$Zr_2Al_3C_4$, 100, 101, 102
$Zr_2Al_3C_5$, 105
$Zr_2Al_4C_5$, 100, 101, 102

$Zr_2AlC$, 99, 100, 101, 102, 103, 104, 106
$Zr_2GeC$, 14, 16
$Zr_2InC$, 44, 45, 47, 48
$Zr_2PbC$, 40
$Zr_2PC$, 100
$Zr_2SC$, 100
$Zr_2SiC$, 100
$Zr_2SnC$, 40, 89
$Zr_3Al_3C_5$, 100, 101, 102
$Zr_3Al_4C_6$, 100, 101, 102
$Zr_3AlC_2$, 100, 102, 103, 105, 106, 107
$Zr_3SnC_2$, 89, 90
$Zr_4Al_3C_6$, 100, 101, 102

# References

[1] Jeitschko, W., Nowotny, H., Benesovsky, F., Monatshefte für Chemie., 94[4], 1963, 672-676. https://doi.org/10.1007/BF00913068

[2] Imtyazuddin, M., Mir, A.H., Tunes, M.A., Vishnyakov, V.M., Journal of Nuclear Materials, 526, 2019, 151742. https://doi.org/10.1016/j.jnucmat.2019.151742

[3] Naveed, M., Obrosov, A., Zak, A., Dudzinski, W., Volinsky, A.A., Weiss, S., Metals, 6[11] 2016, 265. https://doi.org/10.3390/met6110265

[4] Gibson, J.S.K.L., Gonzalez-Julian, J., Krishnan, S., Vassen, R., Korte-Kerzel, S., Journal of the European Ceramic Society, 39[16] 2019,. 5149-5155. https://doi.org/10.1016/j.jeurceramsoc.2019.07.045

[5] Obrosov, A., Gulyaev, R., Zak, A., Ratzke, M., Naveed, M., Dudzinski, W., Weiss, S., Materials, 10[2] 2017, 156. https://doi.org/10.3390/ma10020156

[6] Liu, J., Zuo, X., Wang, Z., Wang, L., Wu, X., Ke, P., Wang, A., Journal of Alloys and Compounds, 753, 2018, 11-17. https://doi.org/10.1016/j.jallcom.2018.04.100

[7] Liu, P., Wang, Z.Y., Ke, P.L., Li, X.W., Wu, X.C., Wang, A.Y., China Surface Engineering, 29[1] 2016, 64-72.

[8] Huang, Q., Han, H., Liu, R., Lei, G., Yan, L., Zhou, J., Huang, Q., Acta Materialia, 110, 2016, 1-7. https://doi.org/10.1016/j.actamat.2016.03.021

[9] Abdelkader, A.M., Journal of the European Ceramic Society, 36[1] 2016, 33-42. https://doi.org/10.1016/j.jeurceramsoc.2015.09.003

[10] Lin, Z., Zhou, Y., Li, M., Wang, J., Materials Research and Advanced Techniques, 96[3] 2005, 291-296. https://doi.org/10.3139/146.101033

[11] Ling, W.D., Wei, P., Zhao, D.Q., Shao, Y.P., Duan, J.Z., Han, J.F., Duan, W.S., Physica B, 552, 2019, 178-183. https://doi.org/10.1016/j.physb.2018.09.041

[12] Li, Y., Zhao, G., Du, B., Li, M., Xu, J., Qian, Y., Sheng, L., Zheng, Y., Surface Innovations, 7[1] 2018, 4-9. https://doi.org/10.1680/jsuin.18.00037

[13] Li, Y., Zhao, G., Qian, Y., Xu, J., Li, M., Journal of Materials Science and Technology, 34[3] 2018, 466-471. https://doi.org/10.1016/j.jmst.2017.01.029

[14] Ying, G., He, X., Li, M., Han, W., He, F., Du, S., Materials Science and Engineering A, 528[6] 2011, 2635-2640. https://doi.org/10.1016/j.msea.2010.12.039

[15] Li, S.B., Yu, W.B., Zhai, H.X., Song, G.M., Sloof, W.G., van der Zwaag, S., Journal of the European Ceramic Society, 31[1-2] 2011, 217-224. https://doi.org/10.1016/j.jeurceramsoc.2010.08.014

[16] Panigrahi, B.B., Chu, M.C., Kim, Y.I., Cho, S.J., Gracio, J.J., Journal of the American Ceramic Society, 93[6] 2010, 1530-1533.

[17] Tian, W., Vanmeensel, K., Wang, P., Zhang, G., Li, Y., Vleugels, J., Van der Biest, O., Materials Letters, 61[22] 2007, 4442-4445. https://doi.org/10.1016/j.matlet.2007.02.023

[18] Tian, W.B., Wang, P.L., Zhang, G.J., Kan, Y.M., Li, Y.X., Journal of the American Ceramic Society, 90[5] 2007, 1663-1666. https://doi.org/10.1111/j.1551-2916.2007.01634.x

[19] Tian, W., Wang, P., Zhang, G., Kan, Y., Li, Y., Yan, D., Materials Science and Engineering A, 454-455, 2007, 132-138. https://doi.org/10.1016/j.msea.2006.11.032

[20] Tian, W.B., Wang, P.L., Zhang, G.J., Kan, Y.M., Li, Y.X., Journal of Inorganic Materials, 22[1] 2007, 189-192.

[21] Han, J.H., Park, S.W., Kim, Y.D., Materials Science Forum, 534-536[2] 2007, 1085-1088. https://doi.org/10.4028/www.scientific.net/MSF.534-536.1085

[22] Yan, M., Duan, X., Zhang, Z., Liao, X., Zhang, X., Qiu, B., Wei, Z., He, P., Rao, J., Zhang, X., Jia, D., Zhou, Y., Journal of the European Ceramic Society, 39[16] 2019, 5140-5148. https://doi.org/10.1016/j.jeurceramsoc.2019.07.052

[23] Grieseler, R., Theska, F., Stürzel, T., Hähnlein, B., Stubenrauch, M., Hopfeld, M., Kups, T., Pezoldt, J., Schaaf, P., Thin Solid Films, 604, 2016, 85-89. https://doi.org/10.1016/j.tsf.2016.03.026

[24] Yu, W., Li, S., Sloof, W.G., Materials Science and Engineering A, 527[21-22] 2010, 5997-6001. https://doi.org/10.1016/j.msea.2010.05.093

[25] Arakia, W., Gonzalez-Julian, J., Malzbender, J., Journal of the European Ceramic Society, 39, 2019, 3660-3667. https://doi.org/10.1016/j.jeurceramsoc.2019.04.047

[26] Berger, O., Boucher, R., Surface Engineering, 33[3] 2017, 192-203. https://doi.org/10.1080/02670844.2016.1201366

[27] Hettinger, J.D., Lofland, S.E., Finkel, P., Meehan, T., Palma, J., Harrell, K., Gupta, S., Ganguly, A., El-Raghy, T., Barsoum, M.W., Physical Review B, 72[11] 2005, 115120. https://doi.org/10.1103/PhysRevB.72.115120

[28] Sun, Z., Li, S., Ahuja, R., Schneider, J.M., Solid State Communications, 129[9] 2004, 589-592. https://doi.org/10.1016/j.ssc.2003.12.008

[29] Amini, S., Zhou, A., Gupta, S., De Villier, A., Finkel, P., Barsoum, M.W., Journal of Materials Research, 23[8] 2008, 2157-2165. https://doi.org/10.1557/JMR.2008.0262

[30] Bouhemadou, A., Applied Physics A, 96[4] 2009, 959-967. https://doi.org/10.1007/s00339-009-5106-5

[31] Liu, Z., Yang, J., Qian, Y., Xu, J., Zuo, J., Li, M., Ceramics International, 46[14] 2020, 22854-22860. https://doi.org/10.1016/j.ceramint.2020.06.055

[32] Aydin, S., Tatar, A., Ciftci, Y.O., Solid State Sciences, 53, 2016, 44-55. https://doi.org/10.1016/j.solidstatesciences.2015.10.010

[33] Roknuzzaman, M., Hadi, M.A., Ali, M.A., Hossain, M.M., Jahan, N., Uddin, M.M., Alarco, J.A., Ostrikov, K., Journal of Alloys and Compounds, 727, 2017, 616-626. https://doi.org/10.1016/j.jallcom.2017.08.151

[34] Chen, H., Yang, L., Physica B, 406[23] 2011, 4489-4493. https://doi.org/10.1016/j.physb.2011.09.013

[35] Sarker, S., Rahman, M.A., Khatun, R., Computational Condensed Matter, 26, 2021, e00512. https://doi.org/10.1016/j.cocom.2020.e00512

[36] Ali, M.A., Hossain, M.M., Islam, A.K.M.A., Naqib, S.H., Journal of Alloys and Compounds, 857, 2021, 158264. https://doi.org/10.1016/j.jallcom.2020.158264

[37] Zhou, Y., Xiang, H., Dai, F., Feng, Z., Journal of the American Ceramic Society, 101[1] 2018, 365-375. https://doi.org/10.1111/jace.15186

[38] Shao, Y., Duan, W., Journal of Applied Physics, 127[15] 2020, 155902. https://doi.org/10.1063/1.5144577

[39] Xu, L., Shi, O., Liu, C., Zhu, D., Grasso, S., Hu, C., Ceramics International, 44[11] 2018, 13396-13401. https://doi.org/10.1016/j.ceramint.2018.04.177

[40] Ling, W.D., Wei, P., Duan, J.Z., Duan, W.S., Modern Physics Letters B, 31[27] 2017, 1750248. https://doi.org/10.1142/S0217984917502487

[41] Hadi, M.A., Computational Materials Science, 117, 2016, 422-427. https://doi.org/10.1016/j.commatsci.2016.02.018

[42] Huang, D., Qiu, R., Mo, C., Fa, T., Computational Materials Science, 137, 2017, 306-313. https://doi.org/10.1016/j.commatsci.2017.05.029

[43] Mebtouche, H., Baraka, O., Yakoubi, A., Khenata, R., Tahir, S.A., Ahmed, R., Naqib, S.H., Bouhemadou, A., Bin Omran, S., Wang, X., Materials Today Communications, 25, 2020, 101420. https://doi.org/10.1016/j.mtcomm.2020.101420

[44] Champagne, A., Ricci, F., Barbier, M., Ouisse, T., Magnin, D., Ryelandt, S., Pardoen, T., Hautier, G., Barsoum, M.W., Charlier, J.C., Physical Review Materials, 4[1] 2020, 013604. https://doi.org/10.1103/PhysRevMaterials.4.013604

[45] Hadi, M.A., Naqib, S.H., Christopoulos, S.R.G., Chroneos, A., Islam, A.K.M.A., Journal of Alloys and Compounds, 724, 2017, 1167-1175. https://doi.org/10.1016/j.jallcom.2017.07.110

[46] Fu, S., Liu, Y., Zhang, H., Grasso, S., Hu, C., Journal of Alloys and Compounds, 815, 2020, 152485. https://doi.org/10.1016/j.jallcom.2019.152485

[47] Hadi, M.A., Ali, M.S., Chinese Physics B, 25[10] 2016, 107103. https://doi.org/10.1088/1674-1056/25/10/107103

[48] Salama, I., El-Raghy, T., Barsoum, M.W., Journal of Alloys and Compounds, 347[1-2] 2002, 271-278. https://doi.org/10.1016/S0925-8388(02)00756-9

[49] Cai, P., He, Q., Wu, X., Liu, X., Liu, Y., Yin, J., Huang, Y., Huang, Z., Ceramics International, 44[16] 2018, 19135-19142. https://doi.org/10.1016/j.ceramint.2018.06.158

[50] Gu, J., Pan, L., Yang, J., Yu, L., Zhang, H., Zou, W., Xu, C., Hu, C., Qiu, T., Journal of the European Ceramic Society, 36[4] 2016, 1001-1008. https://doi.org/10.1016/j.jeurceramsoc.2015.10.023

[51] Hu, C., Sakka, Y., Nishimura, T., Guo, S., Grasso, S., Tanaka, H., Science and Technology of Advanced Materials, 12[4] 2011, 044603. https://doi.org/10.1088/1468-6996/12/4/044603

[52] Hu, C., Sakka, Y., Tanaka, H., Nishimura, T., Grasso, S., Journal of Alloys and Compounds, 487[1-2] 2009, 675-681. https://doi.org/10.1016/j.jallcom.2009.08.036

[53] Zhang, H., Hu, T., Li, Z., Zhang, Y., Hu, M., Wang, X., Zhou, Y., Journal of the American Ceramic Society, 100[2] 2017, 724-731. https://doi.org/10.1111/jace.14559

[54] Hadi, M.A., Kelaidis, N., Naqib, S.H., Islam, A.K.M.A., Chroneos, A., Vovk, R.V., Journal of Physics and Chemistry of Solids, 149, 2021, 109759. https://doi.org/10.1016/j.jpcs.2020.109759

[55] Ding, H., Li, Y., Lu, J., Luo, K., Chen, K., Li, M., Persson, P.O.Å., Hultman, L., Eklund, P., Du, S., Huang, Z., Chai, Z., Wang, H., Huang, P., Huang, Q., Materials Research Letters, 7[12] 2019, 510-516. https://doi.org/10.1080/21663831.2019.1672822

[56] Chen, J.J., Duan, J.Z., Wang, C.L., Duan, W.S., Yang, L., Computational Materials Science, 82, 2014, 521-524. https://doi.org/10.1016/j.commatsci.2013.08.008

[57] Ali, M.S., Parvin, F., Islam, A.K.M.A., Hossain, M.A., Computational Materials Science, 74, 2013, 119-123. https://doi.org/10.1016/j.commatsci.2013.03.020

[58] Shein, I.R., Ivanovskii, A.L., Physica B, 410[1] 2013, 42-48. https://doi.org/10.1016/j.physb.2012.10.036

[59] Bouhemadou, A., Khenata, R., Binomran, S., Physica B, 406[14] 2011, 2851-2857. https://doi.org/10.1016/j.physb.2011.04.047

[60] Li, C., Kuo, J., Wang, B., Li, Y., Wang, R., Journal of Physics D, 42[7] 2009, 075404. https://doi.org/10.1088/0022-3727/42/7/075404

[61] Ali, M.S., Parvin, F., Islam, A.K.M.A., Hossain, M.A., Computational Materials Science, 74, 2013, 119-123. https://doi.org/10.1016/j.commatsci.2013.03.020

[62] El-Raghy, T., Chakraborty, S., Barsoum, M.W., Journal of the European Ceramic Society, 20[14-15] 2000, 2619-2625. https://doi.org/10.1016/S0955-2219(00)00127-8

[63] Barsoum, M.W., Yaroschuck, G., Tyagi, S., Scripta Materialia, 37[10] 1997, 158-1591. https://doi.org/10.1016/S1359-6462(97)00288-1

[64] Lapauw, T., Tytko, D., Vanmeensel, K., Huang, S., Choi, P.P., Raabe, D., Caspi, E.N., Ozeri, O., To Baben, M., Schneider, J.M., Lambrinou, K., Vleugels, J., Inorganic Chemistry, 55[11] 2016, 5445-5452. https://doi.org/10.1021/acs.inorgchem.6b00484

[65] Hossain, M.A., Ali, M.S., Parvin, F., Islam, A.K.M.A., Computational Materials Science, 73, 2013, 1-8. https://doi.org/10.1016/j.commatsci.2013.02.017

[66] Chowdhury, A., Ali, M.A., Hossain, M.M., Uddin, M.M., Naqib, S.H., Islam, A.K.M.A., Physica Status Solidi, 255[3] 2018, 1700235. https://doi.org/10.1002/pssb.201700235

[67] Razzak, M.A., Ali, M.S., Hossain, M.A., Computational Condensed Matter, 13, 2017, 41-48. https://doi.org/10.1016/j.cocom.2017.09.001

[68] Pilemalm, R., Simak, S., Eklund, P., Results in Physics, 13, 2019, 102293. https://doi.org/10.1016/j.rinp.2019.102293

[69] Griseri, M., Tunca, B., Lapauw, T., Huang, S., Popescu, L., Barsoum, M.W., Lambrinou, K., Vleugels, J., Journal of the European Ceramic Society, 39[10] 2019, 2973-2981. https://doi.org/10.1016/j.jeurceramsoc.2019.04.021

[70] Sultana, F., Uddin, M.M., Ali, M.A., Hossain, M.M., Naqib, S.H., Islam, A.K.M.A., Results in Physics, 11, 2018, 869-876. https://doi.org/10.1016/j.rinp.2018.10.044

[71] Lin, Z.J., Zhuo, M.J., Zhou, Y.C., Li, M.S., Wang, J.Y., Acta Materialia, 54[4] 2006, 1009-1015. https://doi.org/10.1016/j.actamat.2005.10.028

[72] Guo, J.M., Chen, K.X., Liu, G.H., Zhou, H.P., Ning, X.S., Journal of Functional Materials, 35[6] 2004, 763-765+768.

[73] Loganathan, A., Sahu, A., Rudolf, C., Zhang, C., Rengifo, S., Laha, T., Boesl, B., Agarwal, A., Surface and Coatings Technology, 334, 2018, 384-393. https://doi.org/10.1016/j.surfcoat.2017.11.067

[74] Kovalev, D.Y., Averichev, O.A., Luginina, M.A., Bazhin, P.M., Russian Journal of Non-Ferrous Metals, 60[1] 2019, 61-67. https://doi.org/10.3103/S1067821219010073

[75] Astapov, I.A., Vlasova, N.M., Ershova, T.B., Kirichenko, E.A., Tsvetnye Metally, 8, 2018, 75-79. https://doi.org/10.17580/tsm.2018.08.10

[76] Xie, X., Yang, R., Cui, Y., Jia, Q., Bai, C., Journal of Materials Science and Technology, 38, 2020, 86-92. https://doi.org/10.1016/j.jmst.2019.05.070

[77] Zhao, Z., Liu, H., Li, X., Zhuang, Y., Zhang, X., Yan, Q., Ceramics International, 46[10] 2020, 14767-14775. https://doi.org/10.1016/j.ceramint.2020.03.001

[78] Haddad, A., Chiker, N., Abdi, M., Benamar, M.E.A., Hadji, M., Barsoum, M.W., Ceramics International, 42[14] 2016, 16325-16331. https://doi.org/10.1016/j.ceramint.2016.07.189

[79] Frodelius, J., Sonestedt, M., Björklund, S., Palmquist, J.P., Stiller, K., Högberg, H., Hultman, L., Surface and Coatings Technology, 202[24] 2008, 5976-5981. https://doi.org/10.1016/j.surfcoat.2008.06.184

[80] Torres, C., Quispe, R., Calderón, N.Z., Eggert, L., Hopfeld, M., Rojas, C., Camargo, M.K., Bund, A., Schaaf, P., Grieseler, R., Applied Surface Science, 537, 2021, 147864. https://doi.org/10.1016/j.apsusc.2020.147864

[81] Haddad, A., Chiker, N., Abdi, M., Benamar, M.E.A., Hadji, M., Barsoum, M.W., Ceramics International, 42[14] 2016, 16325-16331. https://doi.org/10.1016/j.ceramint.2016.07.189

[82] Li, S., Hu, S., Hee, A.C., Zhao, Y., Surface and Coatings Technology, 281, 2015, 164-168. https://doi.org/10.1016/j.surfcoat.2015.09.059

[83] Bai, Y., He, X., Wang, R., Sun, Y., Zhu, C., Wang, S., Chen, G., Journal of the European Ceramic Society, 33[13-14] 2013, 2435-2445. https://doi.org/10.1016/j.jeurceramsoc.2013.04.014

[84] Bai, Y., He, X., Li, Y., Zhu, C., Zhang, S., Journal of Materials Research, 24[8] 2009, 2528-2535. https://doi.org/10.1557/jmr.2009.0327

[85] Wang, X., Zhou, Y., Materials Research and Advanced Techniques, 93[1] 2002, 66-71. https://doi.org/10.3139/146.020066

[86] Guo, J., Wang, B., Chen, K., Zhou, H., Rare Metal Materials and Engineering, 36[5] 2007, 865-868.

[87] Manoun, B., Zhang, F.X., Saxena, S.K., El-Raghy, T., Barsoum, M.W., Journal of Physics and Chemistry of Solids, 67[9-10] 2006, 2091-2094. https://doi.org/10.1016/j.jpcs.2006.05.051

[88] Hettinger, J.D., Lofland, S.E., Finkel, P., Meehan, T., Palma, J., Harrell, K., Gupta, S., Ganguly, A., El-Raghy, T., Barsoum, M.W., Physical Review B, 72[11] 2005, 115120. https://doi.org/10.1103/PhysRevB.72.115120

[89] Radovic, M., Barsoum, M.W., Ganguly, A., Zhen, T., Finkel, P., Kalidindi, S.R., Lara-Curzio, E., Acta Materialia, 54[10] 2006, 2757-2767. https://doi.org/10.1016/j.actamat.2006.02.019

[90] Zhou, W.B., Mei, B.C., Zhu, J.Q., Hong, X.L., Materials Letters, 59[1] 2005, 131-134. https://doi.org/10.1016/j.matlet.2004.07.052

[91] Osada, T., Watabe, A., Yamamoto, J., Brouwer, J.C., Kwakernaak, C., Ozaki, S., van der Zwaag, S., Sloof, W.G., Scientific Reports, 10[1] 2020, 18990. https://doi.org/10.1038/s41598-020-75552-1

[92] Zhan, Z., Chen, Y., Radovic, M., Srivastava, A., Materials Research Letters, 8[7] 2020, 275-281. https://doi.org/10.1080/21663831.2020.1748733

[93] Hashimoto, S., Takeuchi, M., Inoue, K., Honda, S., Awaji, H., Fukuda, K., Zhang, S., Materials Letters, 62[10-11] 2008, 1480-1483. https://doi.org/10.1016/j.matlet.2007.09.005

[94] Singh, J., Wani, M.F., Banday, S., Shekhar, C., Singh, G., IOP Conference Series - Materials Science and Engineering, 561[1] 2019, 012111. https://doi.org/10.1088/1757-899X/561/1/012111

[95] Wada, Y., Sekido, N., Ohmura, T., Yoshimi, K., Materials Transactions, 59[5] 2018, 771-778. https://doi.org/10.2320/matertrans.MBW201703

[96] Wada, Y., Sekido, N., Ohmura, T., Yoshimi, K., Journal of the Japan Institute of Metals, 82[5] 2018, 162-168. https://doi.org/10.2320/jinstmet.J2017042

[97] Liu, S., Wang, C., Yang, T., Fang, Y., Huang, Q., Wang, Y., Nuclear Instruments and Methods in Physics Research B, 406, 2017, 662-669. https://doi.org/10.1016/j.nimb.2017.01.040

[98] Tallman, D.J., He, L., Garcia-Diaz, B.L., Hoffman, E.N., Kohse, G., Sindelar, R.L., Barsoum, M.W., Journal of Nuclear Materials, 468, 2016, 194-206. https://doi.org/10.1016/j.jnucmat.2015.10.030

[99] Wang, C., Yang, T., Xiao, J., Liu, S., Xue, J., Wang, J., Huang, Q., Wang, Y., Acta Materialia, 98, 2015, 197-205. https://doi.org/10.1016/j.actamat.2015.07.043

[100] Bhattacharya, R., Benitez, R., Radovic, M., Goulbourne, N.C., Materials Science and Engineering A, 598, 2014, 319-326. https://doi.org/10.1016/j.msea.2014.01.032

[101] Zhou, A.G., Barsoum, M.W., Basu, S., Kalidindi, S.R., El-Raghy, T., Acta Materialia, 54[6] 2006, 1631-1639. https://doi.org/10.1016/j.actamat.2005.11.035

[102] Du, Y.L., Chinese Physics Letters, 26[11] 2009, 117102. https://doi.org/10.1088/0256-307X/26/11/117102

[103] Gao, L., Han, T., Guo, Z., Zhang, X., Pan, D., Zhou, S., Chen, W., Li, S., Advanced Powder Technology, 31[8] 2020, 3533-3539. https://doi.org/10.1016/j.apt.2020.06.042

[104] Gong, Y., Tian, W., Zhang, P., Chen, J., Zhang, Y., Sun, Z., Journal of Advanced Ceramics, 8[3] 2019, 367-376. https://doi.org/10.1007/s40145-019-0318-4

[105] Prikhna, T., Ostash, O., Basyuk, T., Ivasyshyn, A., Sverdun, V., Loshak, M., Dub, S., Podgurska, V., Moshchil, V., Cabioc, H.T., Chartier, P., Karpets, M., Kovylaev, V.,

Starostina, O., Kozyrev, A., Solid State Phenomena, 230, 2015, 140-143.
https://doi.org/10.4028/www.scientific.net/SSP.230.140

[106] Prikhna, T.A., Starostina, A.V., Lizkendorf, D., Petrusha, I.A., Ivakhnenko, S.A.,
Borimskii, A.I., Filatov, Y.D., Loshak, M.G., Serga, M.A., Tkach, V.N., Turkevich,
V.Z., Sverdun, V.B., Klimenko, S.A., Turkevich, D.V., Dub, S.N., Basyuk, T.V.,
Karpets, M.V., Moshchil, V.E., Kozyrev, A.V., Ilnitskaya, G.D., Kovylyaev, V.V.,
Cabiosh, T., Chartier, P., Journal of Superhard Materials, 36[1] 2014, 9-17.
https://doi.org/10.3103/S106345761401002X

[107] Han, J.H., Hwang, S.S., Lee, D., Park, S.W., Journal of the European Ceramic
Society, 28[5] 2008, 979-988. https://doi.org/10.1016/j.jeurceramsoc.2007.09.015

[108] Wang, C.A., Zhou, A., Peng, C., Huang, Y., Key Engineering Materials, 280-283[2].
2005, 1365-1368. https://doi.org/10.4028/www.scientific.net/KEM.280-283.1365

[109] Rutkowski, P., Huebner, J., Kata, D., Chlubny, L., Lis, J., Witulska, K., Journal of
Thermal Analysis and Calorimetry, 137[6] 2019, 1891-1902.
https://doi.org/10.1007/s10973-019-08107-w

[110] Zhan, Z., Radovic, M., Srivastava, A., Scripta Materialia, 194, 2021, 113698.
https://doi.org/10.1016/j.scriptamat.2020.113698

[111] Wan, D.T., Meng, F.L., Zhou, Y.C., Bao, Y.W., Chen, J.X., Journal of the European
Ceramic Society, 28[3] 2008, 663-669.
https://doi.org/10.1016/j.jeurceramsoc.2007.07.011

[112] Drouelle, E., Joulain, A., Cormier, J., Gauthier-Brunet, V., Villechaise, P., Dubois, S.,
Sallot, P., Journal of Alloys and Compounds, 693, 2017, 622-630.
https://doi.org/10.1016/j.jallcom.2016.09.194

[113] Aryal, S., Sakidja, R., Ouyang, L., Ching, W.Y., Journal of the European Ceramic
Society, 35[12] 2015, 3219-3227. https://doi.org/10.1016/j.jeurceramsoc.2015.03.023

[114] Li, X., Gonzalez-Julian, J., Malzbender, J., Journal of the European Ceramic Society,
40[13] 2020, 4445-4453. https://doi.org/10.1016/j.jeurceramsoc.2020.05.017

[115] Jia, G.Z., Yang, L.J., Physica B, 405, 2010, 4561-4564.
https://doi.org/10.1016/j.physb.2010.08.038

[116] Wenbo, Y., Jia, W., Guo, F., Ma, Z., Zhang, P., Tromas, C., Gauthier-Brunet, V.,
Kent, P.R.C., Sun, W., Dubois, S., Journal of the European Ceramic Society, 40[6]
2020, 2279-2286. https://doi.org/10.1016/j.jeurceramsoc.2020.02.014

[117] Salvo, C., Chicardi, E., García-Garrido, C., Jiménez, J.A., Aguilar, C., Usuba, J., Mangalaraja, R.V., Ceramics International, 45[14] 2019, 17793-17799. https://doi.org/10.1016/j.ceramint.2019.05.350

[118] Zhang, T.F., Xia, Q., Wan, Z., Wang, Q.M., Kim, K.H., Ceramics International, 45[3] 2019, 3940-3947. https://doi.org/10.1016/j.ceramint.2018.11.067

[119] Wang, T., Chen, Z., Wang, G., Wang, L., Zhang, G., Journal of the European Ceramic Society, 38[15] 2018, 4892-4898. https://doi.org/10.1016/j.jeurceramsoc.2018.07.028

[120] Zhang, Z., Jin, H., Chai, J., Pan, J., Seng, H.L., Goh, G.T.W., Wong, L.M., Sullivan, M.B., Wang, S.J., Applied Surface Science, 368, 2016, 88-96. https://doi.org/10.1016/j.apsusc.2016.01.229

[121] Zhang, Z., Jin, H., Chai, J., Shen, L., Seng, H.L., Pan, J., Wong, L.M., Sullivan, M.B., Wang, S.J., Journal of Physical Chemistry C, 118[36] 2014, 20927-20939. https://doi.org/10.1021/jp505428a

[122] Joelsson, T., Flink, A., Birch, J., Hultman, L., Journal of Applied Physics, 102[7] 2007, 074918. https://doi.org/10.1063/1.2785958

[123] Choi, E.S., Sung, J., Wang, Q.M., Kim, K.H., Busnaina, A., Kang, M.C., Transactions of Nonferrous Metals Society of China, 22[S3] 2012, s781-s786. https://doi.org/10.1016/S1003-6326(12)61804-4

[124] Gilev, V.G., Kachenyuk, M.N., Refractories and Industrial Ceramics, 59[6] 2019, 658-662. https://doi.org/10.1007/s11148-019-00291-4

[125] Joelsson, J., Höfling, A., Birch, J., Hultman, L., Applied Physics Letters, 86[11] 2005, 111913. https://doi.org/10.1063/1.1882752

[126] Song, J., Yang, R., Mei, B., Journal of the Chinese Ceramic Society, 39[9] 2011, 1439-1444.

[127] Burka, M.P., Gorban, V.F., Demidik, A.M., Ivanova, I.I., Krylova, N.A., Pechkovs'ky, Eh.P., Polushko, A.P., Samelyuk, A.V., Firstov, S.O., Metallofizika i Noveishie Tekhnologii, 28[6] 2006, 749-768.

[128] Procopio, A.T., Barsoum, M.W., El-Raghy, T., Metallurgical and Materials Transactions A, 31[2] 2000, 333-337. https://doi.org/10.1007/s11661-000-0268-y

[129] Gao, H., Benitez, R., Son, W., Arroyave, R., Radovic, M., Materials Science and Engineering A, 676, 2016, 197-208. https://doi.org/10.1016/j.msea.2016.08.098

[130] Cai, L., Huang, Z., Hu, W., Lei, C., Wo, S., Li, X., Zhai, H., Zhou, Y., Ceramics International, 44[8] 2018, 9593-9600. https://doi.org/10.1016/j.ceramint.2018.02.183

[131] Cai, L., Huang, Z., Hu, W., Lei, C., Wo, S., Li, X., Zhai, H., Zhou, Y., International Journal of Applied Ceramic Technology, 15[5] 2018, 1212-1221. https://doi.org/10.1111/ijac.12902

[132] Ching, W.Y., Mo, Y., Aryal, S., Rulis, P., Journal of the American Ceramic Society, 96[7] 2013, 2292-2297. https://doi.org/10.1111/jace.12376

[133] Phatak, N.A., Saxena, S.K., Fei, Y., Hu, J., Journal of Alloys and Compounds, 474[1-2] 2009, 174-179. https://doi.org/10.1016/j.jallcom.2008.06.073

[134] Ganguly, A., Zhen, T., Barsoum, M.W., Journal of Alloys and Compounds, 376[1-2] 2004, 287-295. https://doi.org/10.1016/j.jallcom.2004.01.011

[135] Han, J.H., Nam, K.D., Park, S.W., Kim, Y.D., Solid State Phenomena, 124-126[1] 2007, 751-754. https://doi.org/10.4028/www.scientific.net/SSP.124-126.751

[136] Finkel, P., Seaman, B., Harrell, K., Hettinger, J.D., Lofland, S.E., Ganguly, A., Barsoum, M.W., Sun, Z., Li, S., Ahuja, R., Physical Review B, 70[8] 2004, 085104. https://doi.org/10.1103/PhysRevB.70.085104

[137] Ali, M.M., Hadi, M.A., Ahmed, I., Haider, A.F.M.Y., Islam, A.K.M.A., Materials Today Communications, 25, 2020, 101600. https://doi.org/10.1016/j.mtcomm.2020.101600

[138] Pan, R., Zhu, J., Liu, Y., Materials Science and Technology, 34[9] 2018, 1064-1069. https://doi.org/10.1080/02670836.2017.1419614

[139] Zhou, W., Liu, L., Zhu, J., Tian, S., Ceramics International, 43[12] 2017, 9363-9368. https://doi.org/10.1016/j.ceramint.2017.04.104

[140] Du, Y.L., Sun, Z.M., Hashimoto, H., Tian, W.B., Physics Letters A, 372[31] 2008, 5220-5223. https://doi.org/10.1016/j.physleta.2008.05.079

[141] Amini, S., Barsoum, M.W., El-Raghy, T., Journal of the American Ceramic Society, 90[12] 2007, 3953-3958.

[142] Du, Y.L., Sun, Z.M., Hashimoto, H., Physica B, 405[2] 2010, 720-723. https://doi.org/10.1016/j.physb.2009.09.093

[143] Bouhemadou, A., Khenata, R., Physics Letters A, 372, 2008, 6448–6452. https://doi.org/10.1016/j.physleta.2008.08.066

[144] Cui, S., Feng, W., Hu, H., Feng, Z., Liu, H., Scripta Materialia, 61, 2009, 576–579. https://doi.org/10.1016/j.scriptamat.2009.05.026

[145] Scabarozi, T.H., Amini, S., Finkel, P., Leaffer, O.D., Spanier, J.E., Barsoum, M.W., Drulis, M., Drulis, H., Tambussi, W.M., Hettinger, J.D., Journal of Applied Physics, 104, 2008, 033502. https://doi.org/10.1063/1.2959738

[146] Xu, X., Ngai, T.L., Li, Y., Ceramics International, 41[6] 2015, 7626-7631. https://doi.org/10.1016/j.ceramint.2015.02.088

[147] Calado, L.D., Padilha, G.S., Osório, W.R., Bortolozo, A.D., International Journal of Advanced Manufacturing Technology, 104[1-4] 2019, 1561-1570. https://doi.org/10.1007/s00170-019-04063-9

[148] Hosseinizadeh, S.A., Pourebrahim, A., Baharvandi, H., Ehsani, N., Ceramics International, 46[14] 2020, 22208-22220. https://doi.org/10.1016/j.ceramint.2020.05.298

[149] Qu, L., Bei, G., Nijemeisland, M., Cao, D., van der Zwaag, S., Sloof, W.G., Ceramics International, 46[2] 2020, 1722-1729. https://doi.org/10.1016/j.ceramint.2019.09.145

[150] Foratirad, H., Maragheh, M.G., Baharvandi, H.R., Materials and Manufacturing Processes, 32[16] 2017, 1874-1880. https://doi.org/10.1080/10426914.2017.1328111

[151] Foratirad, H., Baharvandi, H.R., Maragheh, M.G., Materials Letters, 180, 2016, 219-222. https://doi.org/10.1016/j.matlet.2016.05.181

[152] Huang, Q., Liu, R., Lei, G., Huang, H., Li, J., He, S., Li, D., Yan, L., Zhou, J., Huang, Q., Journal of Nuclear Materials, 465, 2015, 640-647. https://doi.org/10.1016/j.jnucmat.2015.06.056

[153] Emmerlich, J., Högberg, H., Sasvári, S., Persson, P.O.Å., Hultman, L., Palmquist, J.P., Jansson, U., Molina-Aldareguia, J.M., Czigány, Z., Journal of Applied Physics, 96[9] 2004, 4817-4826. https://doi.org/10.1063/1.1790571

[154] Palmquist, J.P., Li, S., Persson, P.O.Å., Emmerlich, J., Wilhelmsson, O., Högberg, H., Katsnelson, M.I., Johansson, B., Ahuja, R., Eriksson, O., Hultman, L., Jansson, U., Physical Review B, 70[16] 2004, 165401. https://doi.org/10.1103/PhysRevB.70.165401

[155] Hu, J.J., Bultman, J.E., Patton, S., Zabinski, J.S., Tribology Letters, 16[1-2] 2004, 113-122. https://doi.org/10.1023/B:TRIL.0000009721.87012.45

[156] Istomin, P., Istomina, E., Nadutkin, A., Grass, V., Leonov, A., Kaplan, M., Presniakov, M., Ceramics International, 43[18] 2017, 16128-16135. https://doi.org/10.1016/j.ceramint.2017.08.180

[157] Nie, J., Liu, S., Zhan, X., Ao, L., Li, L., Physica B, 571, 2019, 105-111. https://doi.org/10.1016/j.physb.2019.06.052

[158] Candan, A., Akbudak, S., Uğur, Ş., Uğur, G., Journal of Alloys and Compounds, 771, 2019, 664-673. https://doi.org/10.1016/j.jallcom.2018.08.286

[159] Wang, X.F., Ma, J.J., Jiao, Z.Y., Zhang, X.Z., Acta Physica Sinica, 65[20] 2016, 206201. https://doi.org/10.7498/aps.65.206201

[160] Bouhemadou, A., Physica B, 403[17] 2008, 2707-2713. https://doi.org/10.1016/j.physb.2008.02.014

[161] Hadi, M.A., Christopoulos, S.R.G., Naqib, S.H., Chroneos, A., Fitzpatrick, M.E., Islam, A.K.M.A., Journal of Alloys and Compounds, 748, 2018, 804-813. https://doi.org/10.1016/j.jallcom.2018.03.182

[162] Zuo, C., Zhong, C., Materials Chemistry and Physics, 250, 2020, 123059. https://doi.org/10.1016/j.matchemphys.2020.123059

[163] Badr, H.O., Champagne, A., Ouisse, T., Charlier, J.C., Barsoum, M.W., Physical Review Materials, 4[8] 2020, 083605. https://doi.org/10.1103/PhysRevMaterials.4.083605

[164] Hossein-Zadeh, M., Mirzaee, O., Mohammadian-Semnani, H., Razavi, M., Ceramics International, 45[18] 2019, 23902-23916. https://doi.org/10.1016/j.ceramint.2019.07.236

[165] Hossein-Zadeh, M., Ghasali, E., Mirzaee, O., Mohammadian-Semnani, H., Alizadeh, M., Orooji, Y., Ebadzadeh, T., Journal of Alloys and Compounds, 795, 2019, 291-303. https://doi.org/10.1016/j.jallcom.2019.05.029

[166] Zhao, G., Ge, F., Cheng, X., Huang, F., China Surface Engineering, 32[3] 2019, 80-87.

[167] Khatun, M.R., Ali, M.A., Parvin, F., Islam, A.K.M.A., Results in Physics, 7, 2017, 3634-3639. https://doi.org/10.1016/j.rinp.2017.09.043

[168] Wang, J., Zhou, Y., Physical Review B, 69[21] 2004, 214111. https://doi.org/10.1103/PhysRevB.69.214111

[169] Cover, M.F., Warschkow, O., Bilek, M.M., McKenzie, D.R., Journal of Physics - Condensed Matter, 21[30] 2009, 305403. https://doi.org/10.1088/0953-8984/21/30/305403

[170] Hu, C., He, L., Liu, M., Wang, X., Wang, J., Li, M., Bao, Y., Zhou, Y., Journal of the American Ceramic Society, 91, 2008, 4029–4035. https://doi.org/10.1111/j.1551-2916.2008.02774.x

[171] Hossein-Zadeh, M., Mirzaee, O., Mohammadian-Semnani, H., Ceramics International, 45[6] 2019, 7446-7457. https://doi.org/10.1016/j.ceramint.2019.01.036

[172] Li, C., Wang, B., Li, Y., Wang, R., Journal of Physics D, 42[6] 2009, 065407.

[173] Wang, J., Zhou, Y., Lin, Z., Hu, J., Scripta Materialia, 58, 2008, 1043–1046. https://doi.org/10.1016/j.scriptamat.2008.01.058

[174] Zhou, Y., Xiang, H., Dai, F.Z., Feng, Z., Journal of the American Ceramic Society, 101[6] 2018, 2459-2470. https://doi.org/10.1111/jace.15398

[175] Ghadimi, M., Baharvandi, H.R., Amadeh, A.A., Materials Letters, 282, 2021, 128831. https://doi.org/10.1016/j.matlet.2020.128831

[176] Ali, M.A., Hossain, M.M., Jahan, N., Islam, A.K.M.A., Naqib, S.H., Computational Materials Science, 131, 2017, 139-145.

[177] Nasir, M.T., Hadi, M.A., Naqib, S.H., Parvin, F., Islam, A.K.M.A., Roknuzzaman, M., Ali, M.S., International Journal of Modern Physics B, 28[32] 2014, 1550022. https://doi.org/10.1142/S0217979215500228

[178] Wang, C., Han, H., Zhao, Y., Zhang, W., Guo, Y., Ren, C., Zeng, G., Huang, Q., Huai, P., Journal of the American Ceramic Society, 101[2] 2018, 756-772. https://doi.org/10.1111/jace.15252

[179] Azzouz-Rached, A., Rached, H., Ouadha, I., Rached, D., Reggad, A., Materials Chemistry and Physics, 260, 2021, 124189. https://doi.org/10.1016/j.matchemphys.2020.124189

[180] Lapauw, T., Halim, J., Lu, J., Cabioc'h, T., Hultman, L., Barsoum, M.W., Lambrinou, K., Vleugels, J., Journal of the European Ceramic Society, 36[3] 2016, 943-947. https://doi.org/10.1016/j.jeurceramsoc.2015.10.011

[181] Wang, J., Zhou, Y., Lin, Z., Liao, T., Physical Review B, 72[5] 2005, 052102. https://doi.org/10.1103/PhysRevB.72.052102

[182] Hadi, M.A., Vovk, R.V., Chroneos, A., Journal of Materials Science: Materials in Electronics, 27[11] 2016, 11925-11933. https://doi.org/10.1007/s10854-016-5338-z

[183] Hadi, M.A., Monira, U., Chroneos, A., Naqib, S.H., Islam, A.K.M.A., Kelaidis, N., Vovk, R.V., Journal of Physics and Chemistry of Solids, 132, 2019, 38-47. https://doi.org/10.1016/j.jpcs.2019.04.010

[184] Zapata-Solvas, E., Hadi, M.A., Horlait, D., Parfitt, D.C., Thibaud, A., Chroneos, A., Lee, W.E., Journal of the American Ceramic Society, 100[8] 2017, 3393-3401. https://doi.org/10.1111/jace.14870

[185] Ouadha, I., Rached, H., Azzouz-Rached, A., Reggad, A., Rached, D., Computational Condensed Matter, 23, 2020, e00468. https://doi.org/10.1016/j.cocom.2020.e00468

www.ingramcontent.com/pod-product-compliance
Lightning Source LLC
Chambersburg PA
CBHW071707210326
41597CB00017B/2366